AI Assisted Business Analytics

Joseph Boffa

AI Assisted Business Analytics

Techniques for Reshaping Competitiveness

 Springer

Joseph Boffa
Boston University
Boston, MA, USA

ISBN 978-3-031-40823-6   ISBN 978-3-031-40821-2   (eBook)
https://doi.org/10.1007/978-3-031-40821-2

This Springer imprint is published by the registered company Springer Nature Switzerland AG
The registered company address is: Gewerbestrasse 11, 6330 Cham, Switzerland

Paper in this product is recyclable.

Author Notes

We are in an evolving business world where large corporation can hire the talent to use analytics for market research, internal statistical audits, data mining, and projecting revenue and expenses. The goal of this book is simple: make statistical analytics an asset for all businesses, especially small business that may have limited budgets. The goal is to avoid discussions that delve immediately into an impenetrable morass of mathematical formulas. Beginner and statistics made easy books are helpful, but what is really needed is a practical way for the busy business manager to have a quick way to get up and running solving practical problems, a classic learn by doing. That is why this book and its companion software are invaluable.

This book includes a complimentary small business/learning version of AuditmetricsAI V6.5 available on the Microsoft Store online. There are no time restrictions. It has the same capabilities as the professional version but with some data and forensic accounting restrictions. This current version was developed for the Massachusetts Department of Revenue Audit Division to get auditors quickly up speed in sales and use tax audits. It is an upgrade of version 5, originally developed for the Massachusetts Division of Health Care Finance and Policy evaluating both private and publicly funded healthcare programs. The upgrade allows the ability to sample accounts containing millions of transactions to meet the specifications of the Massachusetts Department of Revenue.

A video overview of the AuditmetricsAI V6.5 software and to download the free small business/learning version go to the link:

https://auditmetricsai.com/

If you are using your phone to view the video, for a better picture set your phone to allow a horizontal view. In the video you will notice that the AI-assisted software will create Excel documentation. For best performance of the AI Assistance software, it is best to first open Excel and leave it in the background. Then call up AuditmetricsAI.

Before starting the case studies, it would be ideal to also download the "Getting Started Document" and do an initial practice run of the AI Assistance software.

Boston   Joseph Boffa
MA, USA

Contents

About the Author

Joseph Boffa is currently a Board Member of the non-profit Massachusetts Alliance for Retired Americans and administers its Vision Care Discount program for retirees of New England, in conjunction with Vision World and Davis Vision. Dr. Boffa was faculty member and past director of the post-doctoral training program in the Department of Health Policy and Health Services Research, Boston University's Goldman School. He was also Plan Director of a dental benefit for employees of Boston University and Boston Medical Center. He currently teaches a Boston University postdoctoral course: "Healthcare Management and Finance."

Part I
Auditmetrics Overview

Chapter 1
Business Prosperity

1.1 Three Essentials for Small Business to Prosper

- *Stay in touch with cashflow*
- *Manage revenue and expenses*
- *Focus on the consumer*

Stay in Touch with Cashflow—Focus on the areas where cash is being held up, such as inventory, equipment purchases, overextended loan commitments especially short-term loans to fix cash shortfall and accounts receivable. By doing so, a business will be in a position to improve current cash flow. Regular use of statistical analytics can help predict future potential shortfalls. The key to wise management is to anticipate cashflow problems before they overwhelm the system. If the first indication of inadequate liquidity is not being able to meet payroll, it then may make recovery untenable.

How does the statistical audit help? As an example, suppose a small manufacturer could not figure out why bank deposits did not reflect anticipated cashflow. The key to a solution was to obtain a random sample to track in detail business processes in place. What was uncovered were delays in sending or the loss of follow-up of invoices for payment. Also uncovered was that many product orders were just sitting in the truck dock and not being delivered in a timely fashion. This was partly due to lack of system updates of customer contact information. After improvements in administrative procedures, account cashflow began to improve bank deposits.

The key for success was random selection. That assures an unbiased look at what is occurring. All too often a snap decision is made as to the cause of a cashflow shortfall. For example, a quick decision may be made based on a most recent customer contract, a recent bank deposit, or the impression of a colleague or subordinate. Any one of these sources can be biased and lead to misleading conclusions.

J. Boffa, *AI Assisted Business Analytics*,
https://doi.org/10.1007/978-3-031-40821-2_1

Only statistical audits based on systematic unbiased observations will assure a higher percentage of successful outcomes.

Statistical Analytics and Auditmetrics—The key to efficient statistical analytics is to draw random samples from relevant accounts. The use of the Auditmetrics® AI assistance system to guide the drawing of statistical samples is operationally efficient and easy to use. In statistical terms the starting point is to set a targeted margin of error, and the software then guides the user to the drawing of a fully validated and documented random sample.

If you are new to statistical analytics and desire additional background, Appendix I provides a summary of statistical methods broken down into three major categories:

1. *Descriptive Statistics*—is to display and make sense of data using common measures and displays and how it can be organized in databases.
2. *Inferential Statistics*—is to generalize from a sample to a total population that could not be examined in its entirety. It involves the making of decisions under the condition of uncertainty by using probability .
3. *Model Building*—is to develop a model of real-world processes. Statistical theory is used to develop a probabilistic model that best describes the relationship between dependent and independent variables. This provides the business manager with a set of very useful and versatile statistical tools. Model building discussion is what is used in performing the statistical projections used throughout this book.

Pay particular attention to the section on model building. This topic links the audit with forecasting and market research.

Accepted Statistical Standards—The American Institute of Certified Public Accountants statement on the statistical audit states the essential feature of statistical sampling:

1. Sample items should have a known probability of selection, for example, random selection.
2. Results should be evaluated mathematically, in accordance with probability theory.
3. If just one of these requirements is met, like random selection, does not mean that the audit is statistical.
4. It is not a statistical audit if no attempt is made to evaluate sample findings mathematically.

There has been a move by auditors to move away from traditional "block sampling." This method of sampling involves selecting a block of contiguous items from an account. For example, an auditor will ask for all invoices for specific months, i.e., for January, April, and November. AICPA and the IRS do not favor block sampling. The problem is mathematically valid references cannot be made beyond the block examined.

National interest in statistical methods started when the IRS set standards for the statistical audit. For many years the IRS was not in favor of using statistical estimates of account data. However, with the growing complexity of tax filing requirements, such as tax credits, the IRS decided to issue statistical audit directives.

Advantages of the Statistical Audit—Statistical random sampling results in efficient defensible audit outcomes since the random selection process eliminates bias. This produces an objective and defensible result. It also provides for advance estimation of sample size that provides both a defense for the reasonableness of the sample and improves audit efficiency. Such efficiencies are based on universally accepted standards that are clearly stated by the AICPA and the IRS.

- However, determination of sample size calls for good analytical skills and decisions by the auditor.
- Statistical methods in sampling an account provides an estimate of sampling error.
- Sampling error is how far a sample estimation might deviate from the value that could be obtained by a 100% examination of the account.
- Statistical sampling especially with AI assistance saves time and money while maintaining analytic integrity.
- The advance estimate of the required sample size usually results in a smaller sample size than might be arrived at by using a judgmental approach.
- Multiple samples may be combined and evaluated.
- When the audit is based on accepted objective and mathematical standards, it is possible for different auditors to participate independently on the same audit.

Sampling Error Control—Statistical audit involves sampling from computerized systems where key account parameters are known. At each step of the sampling process, key indicators can be monitored to provide immediate feedback as to whether the sample is within accepted standards.

Suppose one wants to sample accounts to find account errors that are not obtainable from standard reports and can only be determined by the detective work of an internal audit:

- Once a random sample is drawn, key statistics can be compared to the parameters derived from a firm's computerized accounting reports.
- For example is a random sample estimate of total dollar value similar to a total determined by standard computer management reports. The term similar has specific statistical analytic meaning.
- Auditmetrics AI automatically goes through validity checks in examining sample statistics in relation to the account parameters sampled. If it encounters a clear mismatch, it will either automatically make an adjustment or alert the user as to what has to be done to rectify the encountered discrepancy. This part of the AI assistance feature assures the design of reliable samples meeting acceptable standards for sampling error.

The terms *statistics* and *parameters* were just used. They have very precise meanings when random samples are used to estimate values of a total book (audit population). For example, in a health insurance medical claims audit, the claims submitted in error in the audit population is unknown. However, a sample can be used to estimate that value. Sample derived statistics can be used to estimate the audit population true value parameter. It would be too costly to measure claim errors directly by examining every claim in the account.

Statistic—Dollars in error based on a sample.

Parameter—Amount of error in the total book which is too costly to measure directly.

The methodology is to use sample statistics to estimate population parameters:

- Statistic ➔ Value derived from a sample
- Parameter ➔ Actual value of the population
- Estimate ➔ Use sample statistics as an unbiased estimate of the parameter

Manage Revenue and Expenses—Once a sample is drawn for an audit, it can be the basis to create datasets to forecast revenue and expenses. The key is to use one of the most powerful statistical tools in building statistical projections, regression, and correlation. Both of these statistical methods are also critical in linking the mathematical relationship between expense inputs and revenue outputs. Only when this relationship is statistically measured can a business consistently manage revenue and cost and forecast future economic performance.

The previous section outlined the essential features of a statistical audit. The Auditmetrics V6.5 systems using AI assistance guides the random sample selection process. The sample is then used to conduct the detective work in determining the number of transactions that have a problem or are in error. Once the accountant acts on the findings of the audit and is assured future cashflow will be well managed, the same sample can be used to project monthly revenue and expenses. In common terms a twofer.

Below is the audit random sample converted into a form to conduct projections of cashflow (Table 1.1):

The regression variables are: **Y — Monthly sales**
X — Month count

Table 1.1 Audit random sample converted into sales by year and month for sales forecast

Month - X	Year	Month	Sales - Y
1	1	January	$129,646
2	1	February	$126,227
3	1	March	$151,966
4	1	April	$165,168
5	1	May	$182,977
6	1	June	$145,694
7	1	July	$156,814
8	1	August	$150,373
9	1	September	$146,155
10	1	October	$202,621
11	1	November	$217,081
12	1	December	$222,458
13	2	January	$198,338
"""	"""	"""	"""

The model to project sales is:

$$Y = a + bX + e$$

Excel built-in functions can be used to calculate the values of intercept (a) and slope (b) that minimizes *residual error (e)*. The residual error of this model is demonstrated in the figure below (Fig. 1.1):

Residual error is a measure of how well the prediction line fits actual account data points. The power of the regression technique is obtained by mathematically fitting the model (straight line) that minimizes the amount of residual error.

Focus on the Consumer—The primary focus of any business has to be an understanding of their customers, their relation to the company, and potential competition. These relationships are at the core of market research. For the manager of a small business, a practical approach is to concentrate on local markets where the business draws its customers. Market research is also a way of getting an overview of consumer wants, needs, and beliefs. It should also involve discovering how the customer perceives the business and relates to it. Market research can be a very expansive discipline which involves the same regression and correlation statistical tools used for forecasting revenue and expenses. These tools are very versatile, and with the help of some built-in Excel functions, any business can set up an effective marketing research plan.

All businesses need information to guide decision-making. Managers trying to understand changing business environments need useable information at the right time. The modern commercial world is inundated with all kinds of data. Data are the collection of facts that are accumulated and stored in account books and customer surveys. That data needs to be converted into useful information so the business manager can make informed decisions.

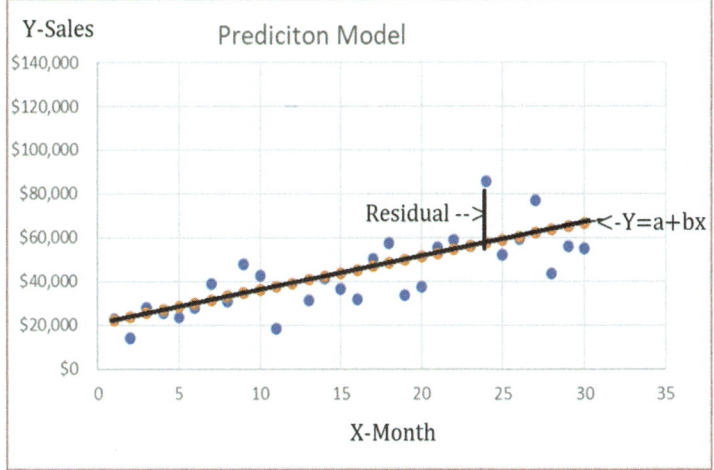

Fig. 1.1 Regression model outlining both prediction line and residual error

Fundamental to the process is to use statistical methods to gain insight of current operations and future needs. All too often business managers may make decisions with assumptions that can be loaded with bias, a bias that is usually based on observations of unstructured customer contact that constrains planning such that one may be following a fluctuation and not a broad trend.

Commercial success more than ever is dependent on business acumen and technological sophistication. Such acumen comes with an understanding of customers' needs. For example, in launching a new product, how will customers react? Will they like it? Will they buy it? How much will they pay? How much will they buy? What will trigger their purchase? Launching a product without this information opens one to basing the decision on hunches and opinion (usually optimistic). Proper collection of customer opinions in conjunction with statistical analytics paves the way for unbiased data-driven decisions.

Commitment to the principles discussed so far will require a learning curve involving both resources and time. No matter how well AuditmetricsAI assistance software eases the transition, it will require improvement of one's comfort with analytic concepts. In terms of customer feedback, a very valuable analytic tool to master is the Likert scale. It is a good entry into the realm of market research.

The Likert scale is in a format in which customer responses are scored along a range. When responding to a Likert item, respondents specify their level of agreement or disagreement on a symmetric agree-disagree scale for a series of statements. Thus, the range captures the intensity of one's feelings for a given item. The Likert scale has found widespread use in business and marketing, primarily because of its simplicity.

The format of a typical five-level Likert item, for example, could be:

- Strongly disagree
- Disagree
- Neither agree nor disagree
- Agree
- Strongly agree

It is a bipolar scaling method, measuring either positive or negative response to a statement. Sometimes an even-point scale is used, where the middle option of "neither agree nor disagree" is not available. This is sometimes called a "forced choice" method, since the neutral option is removed. The neutral option can be seen as an easy option to take when a customer is unsure, but there is much discussion if a true neutral option or just when the respondent is confused. Some researchers claim that there is not a significant difference in the use of forced neutral or not.

1.2   Generic Likert Scale Example

Rating Scale: 1. Very Unsatisfied 2. Unsatisfied 3. Neutral 4. Satisfied 5. Very Satisfied (Fig. 1.2)

Quality of service or Product   1.........2.........3.......4.......5

Price offered   1.........2.........3.......4.......5

Speed of Service   1.........2.........3.......4.......5

Great customer Support   1.........2.........3.......4.......5

Product Warranties   1.........2.........3.......4.......5

Recommend to Friends   1.........2.........3.......4.......5

The Product was Affordable   1.........2.........3.......4.......5

Product is easy to use   1.........2.........3.......4.......5

Recommend product to others   1.........2.........3.......4.......5

Improvements Should be made   1.........2.........3.......4.......5

Staff were prompt In helping me   1.........2.........3.......4.......5

Fig. 1.2 Example of Likert scales to obtain customer's opinions

Table 1.2 Customer ratings of price of a product

Price of Product Rating		
Customer	Product 1	Product 2
1	2	2
2	3	2
3	3	2
4	4	3
5	4	3
6	4	4
7	5	4
8	5	4
9	5	5
10	5	5
Very Unsatisfied	0%	0%
Percent Unsatisfied	10%	30%
Neutral	20%	20%
Satisfied	30%	30%
Very Satisfied	40%	20%

Figure 1.2 is an example of using Likert scale rating system in rating of a specific product in follow-up after a purchase. It can be used to rate several dimensions of customer inputs to obtain a comprehensive impression for each. Excel is ideal in quickly tabulating rankings of each product. Suppose pricing is a major area of concern (Table 1.2):

In essence, the use of averages cannot account for the importance of capturing and understanding variability. At its most fundamental level, the problem is that the numbers in a Likert scale are not numbers as such, but a means of ranking responses. Ordinal scale of measurement can give a rank, but the interval between ranks is unknown. Is the emotional distance between 3 and 4 the same as between 1 and 2?

If the numbers are replaced with the letters A to E, the idea of averaging is meaningless.

The scaling issues listed become paramount when questioning individuals regarding socially or highly personal issues. But for the small business manager, the dimensions are primarily three dimensions of primary concern:

1. Overall customer satisfaction with interaction with the business.
2. Satisfaction of value of goods or services based on price and other perceived values.
3. Likelihood of repeat business and recommend to others.

For the small business, these scales are a simple straightforward indicator of customer satisfaction with the business. They do not deal with more highly charged social and personal assessments. An average to summarize results will yield worthwhile results.

Auditmetrics Sample for Likert Ranking—The audit is conducted to determine deficiencies in managing cashflow. If cashflow is properly managed, then regression forecasts will provide accurate estimates of future trends. Post forecast data collection will determine if new data matches expectations from forecast. Forecasts can serve as an early warning system for detecting potential problems. The Likert rating system is conducted to link cashflow monitoring with the needs and opinions of consumers. The intent is to have customer-linked markers for monitoring and improving a firm's economic outlook.

Auditing and forecasting are great tools up to a point. But to reach full potential, customer feedback other market analytic methods discussed later provides a means to expand it. Customer feedback is essential for businesses to discover what consumers think of their company, product, or service. If companies can successfully identify what their customers value, they can react faster, more flexibly, and more effectively to both changes in the market and customer expectations and needs.

Attribute vs. Variable Sample—But how does one sample accounts for the purpose of obtaining reliable and valid feedback by linking market potential and cashflow? There are two basic types of data sampling techniques, either variable sampling or attribute sampling. When data points are measurements on a quantitative numerical scale, they are variable data, e.g. , weight, length, and, in the statistical audit, dollars. Variable sampling is standard for sales and use tax sampling or sampling in any situation where one wants to measure dollar quantities, such as total dollars in error or in compliance with regulations or other standards. Variable sampling is the standard for the audit of income and sales tax transactions. The Internal Revenue Service (IRS) has established statistical guidelines when projecting revenues and/or expenses from a sample. Variable sampling techniques are driven by quantitative mathematical projections. The fundamental goal of variable sampling is to answer the question of quantity of dollars involved. For example, how much do I owe in taxes or what is the value of my tax credit?

Attribute sampling data are classified categorically. For example, tracking whether accounts receivable items are past due could be categorized as "yes" or "no." Attribute sampling is integral to opinion polling and market research where the pollster seeks the characteristics of targeted subsets of the population. In that environment, the data stratifications are situation dependent. For example, a market researcher may be interested in demographic breakdowns of potential customers such as socioeconomic status and gender. There are multiple variations of attributes, but to the accountant, they should ultimately relate to that all-important variable, dollars.

Attribute sampling is classification dependent. For example, an accountant may want to examine customer base in terms of sex, age category, or other sociodemographic classifications. These attributes are defined such that they are mutually exclusive. The purpose of the analysis would be to answer the questions such as "how many purchasers of a specific good or service are less than age twentyone?"

Just as the Auditmetrics-derived random sample is used to conduct an audit and then forecast revenue and expenses, it can also be used to monitor customer attitudes. The key is to link customer feedback with dollar impact (Table 1.3).

Financial data tends to be skewed to the right which means the vast majority of the customer invoices will cluster in the lower dollar strata, especially compared to those in the upper dollar strata. For example, the first stratum 0–49.99 accounts for 47% of the number of customer invoices but with only 5% of the dollar volume, while the two highest dollar strata account for 44% of dollar volume.

In measuring customer attitudes, selecting those in the upper strata should be given higher preference since they are the group that provides the greatest source of revenue. But there is value in examining those in the lower to middle strata. One of the pioneers of statistics, R. A. Fisher, emphasized the importance of taking into account variability. It is diverse differences that add insight into the dynamics a business' cashflow. Obtaining feedback from those in the lower strata will provide a broader view that can help gain wider insight into cashflow including possible leads for future growth.

Table 1.3 Cashflow of customer invoices broken down by dollar strata

Audit Population				
Strata	Acct.Total Value	Precent Dollars	Invoice Count	Percent Count
0 -49.99	$192,150	5%	10,101	47%
50-174.99	$529,492	13%	5,776	27%
175-399.99	$727,690	17%	2,813	13%
400-824.99	$885,532	21%	1,614	7%
825-1600	$982,161	23%	888	4%
> 1600	$870,549	21%	463	2%
	$4,187,575	100%	21,655	100%

Chapter 2
Analytics Case Studies

Case Study 1 Managing Cashflow—Before starting this case study, it would be helpful to view a 4-min video that demonstrates the interconnected parts of the case study's analysis from initial audit to final financial projection. The background of the case study is that an auditor encountered a decrease in cashflow that required attention. The first issue was which accounts needed to be sampled for a statistical audit. Once decided, a random sample with AI assistance is selected. The sample selection is where the video starts. The video can be viewed at:

https://auditmetricsai.com/

If you are using your phone to view the video, for a better picture set your phone to allow a horizontal view. For those who use Microsoft Windows, there is available at the same site a link to the free Small Business/Learning Auditmetrics V6.5. Also Available on another link is the document "Getting Started" which covers how to set up and run the software. With viewing the video and following the book's discussion, one can replicate in real time the case study. This gives the reader the ability to follow the case study analytics as it would be handled in a real-world business setting, a classic learn-by-doing experience.

Auditmetrics had to pass a Microsoft certification process to qualify for the Universal Windows Platform (UWP). Auditmetrics is in the Desktop and Xbox section of UWP. There are some Microsoft computers that are mini-versions with reduced capabilities. They are not for large business applications, but for those, especially students, who value cost savings and want to learn Auditmetrics, they will work fine.

However, if problems do arise contact *support@auditmetrics.com* for help. If necessary, we can provide the free learning software directly. The UWP protocol is written on top of what is called the .Net (dot net) Framework. The .NET Framework is a standardized software framework developed by Microsoft that runs primarily on Microsoft Windows but can be used to create both desktop-based and Web-based applications.

J. Boffa, *AI Assisted Business Analytics*,
https://doi.org/10.1007/978-3-031-40821-2_2

Before the Microsoft store downloads Auditmetrics, your computer may be updated with the latest .NET Framework. Auditmetrics software managers keep up to date with .NET.

Managing Cashflow—A business should always be familiar with the present state of its cashflow. Just because there is a positive balance in the checking account doesn't mean it has sufficient available cash. The common rule of thumb is for businesses to have a cash buffer of 3–6 months' worth of operating expenses. However, this amount can depend on many factors such as the industry, what stage the business is in, its goals, and access to funding.

There are three types of cashflow:

1. *Operating activities*

   Operating cashflow is a measure of the amount of cash generated by a company's normal business operations. Operating cashflow indicates whether a company can generate sufficient positive cashflow to maintain and grow its operations; otherwise, it may require external financing for capital expansion.

2. *Investment activities*

   Cashflow from investing activities includes any inflows or outflows of cash from a company's long-term investments. The cashflow statement reports the amount of cash and cash equivalents leaving and entering a company.

3. *Financing activities*

   Cashflow from financing activities is the net amount of funding a company generates in a given time period. Receiving cash from issuing stock or spending cash to repurchase shares. Receiving cash from issuing debt or paying down debt.

   Paying cash dividends to shareholders. Proceeds received from employees exercising stock options.

Auditmetrics is designed for businesses to manage cashflow derived from the ongoing operations of the business. It estimates the money expected to flow in and out including all income and expenses. Typically, most business cashflow projections cover a 12-month period. However, businesses can create a weekly, monthly, or semi-annual cashflow projection. Operating cashflow is the foundation for the managing of long-term investment and financing activities.

Financial Information Control—There are many computerized business management systems that provide an array of reports broken down by multiple factors. They are snapshots of current, past, and cumulative account activity which are of great value in measuring business performance. The value of the random sampling system in conjunction with regression forecasting techniques provides expanded capabilities. Our economy is on a high-tech commercial revolution where huge business entities marshal vast sophisticated programming to correct past deficiencies and search for potential opportunities. This book's software helps the smaller enterprise survive by providing it a level analytical playing field.

Random sampling ensures that analytical results will approximate what would have been obtained if the entire account had been measured. Random sampling allows all transactions in an account to have an equal chance of being selected. The sample also allows an auditor to make strong statistical inferences including measuring and quantifying sampling risk (statistical error). Besides statistical inference, the same sample can be the basis for using regression statistical methods for forecasting and model building.

For instance, in this case study, a business manager noticed cashflow restrictions in covering expenses. The decision was to conduct an internal audit to uncover the source of the problem. The first step is to decide which accounts should be given scrutiny and then to draw a random sample and pull invoices to do the necessary detective work. What follows in the case study is structuring the audit to quantify audit recommendations.

In summary the audit process is:

1. Sample selection.
2. Generating Excel templates for documenting audit results.
3. Implement the necessary actions for recovery.

With the task of accomplishing the goal of improving operational cashflow, the next step is to put in place statistical tools to manage long-term dynamics of cashflow. This is where regression modeling can offer wider array of cashflow controls.

The previously discussed prediction model was a good start:

$$\textbf{Monthly Sales} = \textbf{a} + \textbf{b}\left(\textbf{Month Count}\right)$$

This basic model can be expanded to adjust for seasonal fluctuations and to determine customer base geographic distribution. In schematic form:

$$\textbf{Monthly Sales} = \textbf{a} + \textbf{b1}\left(\textbf{Month Count}\right) + \textbf{b2}\left(\textbf{Quarter}\right) + \textbf{b3}\left(\textbf{Geographic Area}\right)^{*}$$

The mathematical concepts and calculations will be covered in Part III

With this basic model, one can now do such things as project sales for the last quarter (October–December) for specific different geographic areas. Geographic area can be based on sales by zip code areas. Most business accounting systems can provide sales by zip code which can be aggregated into meaningful geographic breakdowns. Both seasonal and geographic adjustments can provide an assessment of sales outlining the business' local market area. The next step would be to link forecasts with customer Likert scale inputs providing the initiation of in-house market research.

2.1   Causes of Cashflow Problems

1. *Not Creating a Cashflow Budget*

A cashflow budget is an estimate of cash you expect to receive and cash you expect to pay payout. Just a routine review of checking account balances doesn't assure readily available cash. Sound proper management can be started by creating a cashflow budget on a monthly or quarterly basis depending on business volume. Cash should also be monitored long-term using regression forecasting techniques. This forward-looking process can help tackle issues while they are easily manageable. This process is a valuable adjunct to the standard annual cost and revenue budget review searching for variances.

2. *Making Payment Errors*

Businesses frequently allow small financial administrative errors to creep in. Carelessness easily leads to pay the wrong amount to the wrong account or to pay the same bill twice. Simple errors like this happen all too often and can be a huge headache when it comes to measuring cashflow. The best preventive measure for these errors is to have a quick and easy way to conduct internal audits. Auditmetrics can adjust sample size by adjusting the precision input. Planned routine quick and inexpensive small audits can quickly detect cashflow problems.

3. *Getting Behind on Billing and Collections*

Making sure customers pay on time is extremely important. A business must have a clear payment policy. Communicate that policy to all employees issuing credit and to customers so that they are aware of the consequences involved in late payments. Sending invoices late encourages customers to pay late. Invoices should be sent to customer as soon as products are sold or services are delivered. This is the fundamental accrual accounting concept of expense and revenue realization, an area where regular statistical audits are of great help.

4. *Offering Credit Without Checking Backgrounds*

Before offering sizable credit to new customers, do a check credit. A quick search will help determine whether or not to extend credit. If their credit or payment history is poor, either don't offer credit or outline very specific payment terms and penalties for not paying on time.

5. *Poorly Managing Inventory*

You can usually receive discounts for buying in bulk from vendors or warehouse stores, but it's not worth it if it means a warehouse full of supplies and unsold items. There is what is called the 80/20 inventory management rule. The 80/20 rule states that 80% of results come from 20% of efforts, customers, or another unit of measurement. When applied to inventory, the rule suggests that companies earn roughly 80% of their profits from 20% of their products. Buying in bulk will make sense for the 20% of the products. Indiscriminate buying in bulk can limit cashflow flexibility. It is interesting how pervasive is the 80/20 rule. Auditmetrics use of statistical monitoring of health insurance claims has found that 80% of claims costs are attributable to 20% of insurance plan members.

6. *Poorly Managing Labor Costs*

Labor cost control means control over the cost incurred on labor. Control over labor costs does not imply forcing low wages per worker or reduce drastically the number of employees. The aim is productivity and keep wages per unit of output as low as possible. Worker productivity can be enhanced by investing in more efficient equipment and operational processes.

7. *Focusing on Profit Instead of Cashflow*

When people think of a successful business, maximizing profit is what comes to mind. This can lead to a mindset to seek that contract with the largest profit margin. However, a variety of smaller jobs with less profit may be a better option. Diversifying clients increases the likelihood of being paid regularly. A single large client that can't pay on time may lead to not having enough to pay supply vendors.

Starting with the Income Statement—The crucial account document in managing cashflow is the income statement. The income statement shows a company's revenues, expenses, and profitability over a period of time. It is also sometimes called a profit-and-loss (P&L) statement or an earnings statement. It shows revenue from selling products or services and expenses to generate that revenue.

The business manager in this case study examined the income statement provided by the company's accounting firm. The income statement tracks revenue and expenses over an annual accounting period. In accrual accounting methods, revenue and expenses are recognized when the good or service is delivered. However, revenue is not the same as cash inflow. Total sales revenue of $2,500.000 indicates this amount of product delivered to customers. Allowing for return refunds, the actual revenue is reduced to $2,470,000. Not all of the revenue is in the form of cash; some of it is tied up in accounts receivable yet is taxable as income. Therefore, the total cash inflow is $2,220,000.

The issue for the business is whether the $250,000 accounts receivable contributes to the cashflow problem. There are measures such as "Receivables Turn Over Ratio" and "Receivable Turnover in Days" that are used to judge if accounts receivables are contributing to a loss of cash. It was discovered that this business' Receivable Turnover in days turned out to be 50.59 days. If the business maintains a policy for payments on credit is 30 days, the receivable turnover in days calculated above indicates that the average customer makes late payments.

Indicted in Table 2.1, expenses totaling $823,000 contribute to cash outflow. But another source of cash outflow is loan principal payments. Loan interest payments are tax deductible expenses, but principal payments are not. Principal payments are cash outflows based on past long-term investment and financing decisions. The total cash outflow is $823,000 plus $50,000 equaling $873,000 leading to net cash outflow before taxes of $1,347,000, but the taxable income is $1,647,000.

The breakdown of Tables 2.1 and 2.2 is the critical first step in developing the cashflow budget. The ultimate purpose of the statistical audit is to keep reported taxable income and cashflow as close as possible.

Table 2.2 Outlines the derivation of net income and cashflow after taxes. The manager noticed that the bad debt write-off of $54,300 represents 22% of the accrued accounts receivable. Late payments and 22% uncollectible write-off point to a constant draining of revenue sources. The rule of thumb for small business is to set aside 30% of income after deductions to cover federal and state taxes.

Setting Concise Audit Objectives—At this stage it is clear that this business is suffering from being "income rich but cash poor." The cashflow breakdown points to the fact there is a problem with the collection of account receivables. A statistical

Table 2.1 Income statement outlining income vs cashflow

Revenue		
Sales	2,500,000	
Less: Adjustments /Refunds	30,000	
Net Revenue		$ 2,470,000
accrued in accounts Rec.	250,000	
Total Cash In-Flow		*$ 2,220,000*
Expenses		
Insurance Expense	50,000	
Heat, Light and Water Expenses	10,800	
Miscellaneous	7,200	
Interest Expense	25,000	
Materials Expense	40,000	
Office Salaries Expense	600,000	
Rent Expense	90,000	
Total Expenses		$ 823,000
Loan Payment (principal)	50,000	
Total Cash Out-Flow		*873,000*
Net Cash Flow Before Taxes		**$ 1,347,000**
Gross Income Before Taxes		$ 1,647,000

Table 2.2 Net cashflow and net income after taxes

Net Cash Flow before Taxes		**$ 1,347,000**
Gross Income Before Taxes		$ 1,647,000
Depreciation - Equipment	100,000	
Bad Debts Expenses	54,300	
Total Book Cost		$ 154,300
Net Taxable Income		$ 1,492,700
Taxes owed (30%)		$ 447,810
Net Income After Taxes		$ 1,044,890
Net Cash Flow After Income Tax		**$ 899,190**

audit would help document the moving parts of the problem. But before jumping to selecting a random sample, careful audit questionnaire design comes first. A question needs to collect concise information that is readily converted into useful data. Imprecise objectives will lead to imprecise questions leading to imprecise results. The ultimate goal is statistical precision fueled by precise data.

Time should be taken in the planning phase to consider the audit objectives that ensure proper criteria and audit steps allowing for delivering on audit objectives. Clearly defined objectives improve the efficiency, effectiveness, and scope of the audit and provide a quantified, defensible financial impact statement. A precise objective is critical to delivering on quality, value-added audit decisions.

An audit objective "to improve the operations of account receivable" is too vague and difficult to answer with a "yes" or "no." A more appropriate audit objective would be: "The accounts receivable process supports the timely and proper management of reviewed invoices." Now this is something one can conclude Yes or No. Were the controls followed? Are they adequate and effective? or Are the controls not adequate, ineffective, or not followed? With properly defined objectives, one will have a better idea of how analytics can be used to support the audit.

2.2   Criteria One—Timeliness

- Examine the payment terms and invoice date, and identify late/early payments.
- On average, an acceptable time line for collecting accounts receivables should not be more than one third longer than the credit period, for example, allowing customers to pay within 30 days but allow up to 40 days.
- Account entries greater than 60 days can be concluded as non-compliant with timeliness criterion.
- Are the reviewed invoices both timely and with a clearly documented trailer record? Can conclude with yes or no. If no, then reviewed invoice is not compliant.

2.3   Criteria Two: Are Official Standards Followed?

- Was the approved well-defined A/R process document followed?
- Were adequately trained staff processing the invoices that led to A/R?
- Can documentation recording standards include customer signatures and dates verified?
- Can tracing of A/R invoice with account ledgers be carried out?
- Does the manager of A/R follow through with clear standards of amount of credit to extend, reasonable consideration of risk, the value of the service, and the customer's ability to pay including past performance?
- In medical offices is there a clear process for linking insurance coverage claims and copayments?

Post Audit Documentation—At this juncture for those who have access to Auditmetrics V6.5 Small Business/Learning Version and the practice data file can run the data from an actual internal audit and generate the Excel audit reports. The data below is from an actual account receivable audit. The Excel Workbook contains a spreadsheet with the audit report as seen on Table 2.3. A second spreadsheet contains the random sample data that was used to record invoice errors. The Workbook can be obtained by entering onto your web browser:

https://auditmetricsai.com/filelibrary/fullaudit.xlsx

Table 2.3 examines what proportion of account dollars were audited and what proportion was in error (failed at proper and timely processing) and where the business fails in managing the A/R invoices. Also included in the Excel Workbook is the data file that generated this report. Look at the data and you will see which invoices were in error or non-compliant. With this feedback the next phase of the audit process takes over, tracing back to the people, processes, and accounts impacted.

The software generates this spreadsheet template with only the column "Amount Error" blank for the auditor to fill in. The generated spreadsheet is imbedded with the Excel functions to calculate the ratio estimation method to project the total dollars in error. The error rate is 4.3% Ratio estimation is considered the method least susceptible to statistical bias. This topic will be expanded in Chap. 4.

The Audit Report and Data Integrity—Traditionally the internal audit was considered an early warning system for business managers to search for things that were going wrong and report them. This is probably why the audit and auditor were generally viewed as being a cop and having a "gotcha" attitude. In this context, the audit can be a source of stress for employees. There has been a change in attitude about internal audits when the Institute of Internal Auditors endorsed the use of the audit to identify weakness in business controls. In this context the statistical audit searches for operational cashflow deficiencies. The new evolving role for the audit should be as a transition to identify and assess the risk to the achievement of business objectives. The auditor should be part of the management team to help shape organization objectives in terms of clear and concise objectives as discussed previously. It is key that objectives should be readily measurable for data analysis.

Table 2.3 Post audit documentation

Acme Inc.	Population		Sample		Audit Results		
Strata	Account Total	Acct. Freq. N	Sample Acct. Total	Sample Size n	Amount Error	Error Ratio	Tot. Est. Error Amt.
0-49.99	$192,150	10,101	$1,277	67	$ 88.22	0.069	$13,272.69
50-174.99	$529,492	5,776	$11,895	124	$ 486.20	0.041	$21,643.30
175-399.99	$727,690	2,813	$36,491	138	$ 1,803.27	0.049	$35,960.40
400-824.99	$885,532	1,614	$90,144	161	$ 4,406.45	0.049	$43,287.01
825-1600	$982,161	888	$222,389	205	$ 9,025.15	0.041	$39,858.71
>1600 (Detail)	$870,549	463	$870,549	463	$ 26,976.44	0.031	$26,976.44
	$4,187,575				Estimated Dollars in Error-->		$180,998.55

Standard business computerized management reports identifying financial integrity issues is a valuable finding. But if management finds that there is incomplete, inaccurate, obsolete, or conflicting data, they will now question the confidence of their data-based decision making. Management and auditor preplanning ensures that both have a clear understanding of the objectives of the audit. Data integrity requires that data be not only accurate but also complete, consistent, and in context. Data integrity is what assures useful audit results.

Completeness—A data record, such as accounts payable, must be complete to satisfy the needs of all its customers. Accounts payable entries must be the correct amount and timely. Gaps in the account's data can hinder an organization's ability to properly manage cash outflow.

Uniqueness—No duplicates. Account entries should not be recorded twice. All too often duplicate records will infect the data which will impact analysis and management decisions.

Timeliness—Data entry should be made as soon as possible after the event being recorded happens. Vendor invoices sitting on a manager's desk for months before being approved and sent to A/P. This hinders vendor relations that may impede the goal of well-managed and coordinated delivery of products and services. Invoices not entered into the A/P system for weeks and weeks can result in under or overpayments.

Accuracy—Wrong or misleading data helps no one. The cause of inaccuracy has many factors including manual input errors, interface or mishandled conflicting data between sources, processing errors, as well as from ineffective analysis that miss or double count records. The result is inaccurate reporting.

Validity—Refers to the accuracy of a measure (whether the results represent what they are supposed to measure). If you only have a single source of data, then it is likely to be consistent, but could be consistently wrong. To truly verify the data, it must be checked against its source, such as a vendor invoice.

Availability/Accessibility—Accurate data must be provided to the proper functions at the right time when it is required. If management needs data today to make a decision, providing the data next week is not useful. A lack of proper and easy access and retrieval is detrimental to the business. It can result in delayed or inaccurate decisions. Further, the data must be accessible in a format that allows it to be used for its ultimate purpose. Proving a printed report with thousands of pages does not support additional drilldown and analysis. The standard Auditmetrics Excel templates are a good start because they are immediately available post-audit.

Consistency/Reliability—Refers to whether the results can be reproduced under the same conditions. If you apply the same data filter or perform the same analysis, you should get the same results. However, consistent data also needs to be valid to be of use. Consistently retrieving 6.500 is useful, but if the actual data is 65.00, then we have a consistency and validity issue.

Context—Data may be accurate, reliable, and accessible, but without proper context, it lacks meaning. Does "555-2381" represent a phone number, an address (suite and street number), or something else? Without context data is just bits and bytes, not useful information.

Case Study 2: Medical Claims Audit Case Study—For several years Auditmetrics has been involved in the design and analysis of healthcare plans. Its initial work was with Taft-Hartley labor/management health and welfare funds designing vision, dental, and pharmacy benefits. Initially its analysis was to analyze the actuarial experience of submitted health insurance claims.

It was later used to do statistical analysis for the Massachusetts Rate Setting Commission which later morphed into the Division of Healthcare Finance and Policy. Auditmetrics was then used to analyze medical claims from Medicaid and all of the State's major health insurers including HMOs and commercial carriers. Teams of auditors and statisticians then used the analysis to establish reimbursement policies for both private insurance and the Medicaid program.

Around the same time, the IRS published directives in setting statistical audit standards, and the Federal Government also set standards for auditing Medicare claims. When Congress passed the Medicare Prescription Drug, Improvement, and Modernization Act, it also set up the Recovery Audit Contractor, or RAC, program to identify and recover improper Medicare payments paid to healthcare providers under fee-for-service Medicare plans. Before this Act, statistical audits of health claims were largely a combination of national and regional standards. Now there is evolving national standards emanating from one of the nation's largest health insurer, Medicare.

The two primary ways through which RACs identify claim error are "automated review" and "complex review." Automated review occurs when an RAC makes a claim determination without a human review of the medical record. RACs use proprietary software designed to detect certain types of errors. For complex reviews which no written Medicare policy/articles/coding exists, a review of the medical record is involved. In those reviews, the RAC must use appropriate medical literature and accepted statistical standards. The RAC's medical director must be involved in actively examining the evidence used in making individual claims determinations.

Medicare also defines an automated review. It is similar to the Auditmetrics automated process to identify medical insurance claims errors such as:

- Incorrect submitted reimbursement amounts
- Prior authorization procedures not properly documented
- Improper coordination of benefits of separate member and spouse plans
- Non-covered services (including services not reasonably necessary)
- Incorrectly coded services
- Duplicate services
- Paid claims with member eligibility errors

The Statistical Audit Model Online—In order to use Auditmetrics to complete this case study, a cut-down Internet version is available at the following website:

https://healthe-link.com/audit/eaudit.aspx

In this exercise

1. *Link to the website. The starting point is to select the "Potential Detail" tab to determine a reasonable detail cutoff. A suggested cutoff of 850 will be displayed on the screen.*
2. *Run with number strata and precision inputs of 5 and 3%.*
3. *Change precision from 3% to 4% to measure impact on sample size of changing precision to 4%.*
4. *To accomplish these steps follow the documentation and steps listed below*

This Internet version software has a streamlined reduced functionality of the full version available from the Microsoft Store. Both versions apply the same assisted artificial intelligence to expedite the statistical audit sampling process. This Internet version generates a random sample. It does not have the Window's version to generate Excel template documentation. Unfortunately, the code to generate Excel files only works on the Windows operating system due to Microsoft copyright restrictions. However, the statistical algorithms are written in the universal .Net (dot net) framework code which means they can run on any Internet server that runs .Net. which are most Internet providers. That is why this case study can be used to provide insight as to the basic functionality of the Auditmetrics system which would include Apple computer users.

For this case study, one can test different input options: detail cutoff, number of strata, and precision (margin of error). They each have a varying impact on sample size and statistical efficiency. In statistical terms the account to be sampled is referred to as the audit population. For this case study, sampling documentation will be displayed in screens #1 and #2. The screen output columns are tab delimited. That means columns may not quite align on the screen, but if you copy and paste them into an Excel spreadsheet, the columns will line up correctly. This will allow one to manipulate and analyze what was displayed on screen.

Detail Strata—This is the strata one does not rely on a sample but 100% review of its invoices. It represents the largest transactions that are eliminated from sampling. This reduces the variability (standard error) of the remaining transactions from which a sample will be drawn. This enhances statistical efficiency. Auditmetrics uses a distribution-based criterion. It starts with examining transaction sizes that range from the 90th percentile to the 99.5th percentile and using AI assistance and selects and displays a suggested detail cutoff.

The detail stratum provides statistical and economic efficiency. In that stratum, the largest transactions of the account are reviewed at 100%. This can represent approximately 25–35% of the account's total dollar volume. The review of these largest transactions is not subject to statistical error. It is an effective use of auditor time by directly examining transactions with the largest economic impact. Determining the detail cutoff is the first step.

The dataset available on the website is of actual medical claims data with a known error rate of 10.2%. To generate the detail cutoff, select the button as indicated below:

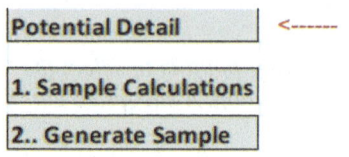

The AI-assisted derived detail cutoff is at $850. Number of strata and precision are the other variables needed to be chosen for input.

Stratification—The next sample input factor that relates to statistical efficiency is dividing the audit population into a number of strata. Random samples are selected from each non-detail stratum. It is both detail cutoff and number of strata that determine overall statistical efficiency. One can vary cutoff, number strata, and precision to determine their effect on sample efficiency. Sampling efficiency is the ratio of the variability (standard error) of the total audit population not stratified as compared to the variability of a stratified audit population. Stratification improves statistical efficiency. For example, a sample efficiency of 84% means that a stratified sample including the detail stratum has 84% less variability or lower standard error than a simple random sample of a non-stratified audit population. When selecting the number of strata, one should find the minimum number of strata that balances both efficiency and sample size. For example, if 5 strata and 6 strata have approximately the same efficiency and sample size, then condense to the more compact lower value of 5.

Stratifying the account is accomplished by using the dollar amount of the transaction as the basis of stratification. Strata boundaries are separated by specific transaction dollar values into a collection of mutually exclusive categories. Auditmetrics AI uses the most efficient methods of determining strata boundaries. It uses the *Cumulative of the square root of the frequency method.* The recommended number of strata can vary from 3 to 10 with 5 in this internet version being the default.

Precision—The most important factor in determining sample size is precision, or as termed by pollsters, "margin of error." The usual way of expressing it is to determine the sample size that would provide sample estimates within a specified range of the true population value. Suppose we have an account book total of $1000. The goal is to have a sample estimate within 3% (precision) of the true population value, which would be in the range of $970 and $1030. Precision and another factor called confidence level are generally used together in estimating population values. A more accurate way of expressing the relationship between precision and confidence is by stating one is 95% confident that sample estimates would be within 3% of the true population value. In statistics 95% confidence means that 95 out of 100 random samples would fall within the 3% precision criterion.

A precision of 3% would require a larger sample size than the wider precision range of 4%. As sample size increases so does statistical power and efficiency. But there is a trade-off in that the increase in the sample size may increase efficiency but offset with increased cost to conduct the audit.

Final Design and Efficiency Factor—Auditmetrics is an expedited way to understand the interplay of the various inputs in determining sample size and statistical efficiency. There are some rules of thumb to follow in assessing statistical efficiency. An acceptable efficiency should be ≥ 0.70. Efficiency <0.70 but >0.60 usually may indicate a highly skewed population. To reduce the effect of extreme skewness, decrease the detail cutoff to a lower the boundary for the sample strata. This will result in reviewing more large volume transactions at 100%. If an efficiency factor does not reach 0.60, the guidance of a statistician should be considered. It does not occur often, but when this has been observed, it usually indicates the audit population is bimodal. Bimodality means that two independent processes or populations are artificially combined into one. The recommended statistical solution is to separate the two populations and audit each separately.

Calculate a sample at the suggested detail cutoff at $850. This results in the number of detail strata transactions to be reviewed at 284. If the precision is 4%, and number of strata to be sampled is 5, then the efficiency is 0.84. If you want to improve efficiency, then include more transactions into your detail. But there is a trade-off with increased cost (Table 2.4):

Including more in the detail stratum increases the number 100% transactions to be selected. The total number of transactions to be reviewed increases to 785 from 763. This increases the cost of doing the audit. There always is an efficiency/cost trade-off. In this example the difference is not that large, but this file was chosen for this discussion because it was one that was not highly skewed. With most financial data, the trade-offs are more pronounced.

Validity Check—Statistical sample audits have an advantage over other types of sampling environments. When health researchers test different models of disease prevention or treatment effectiveness, a statistical sample is used to estimate a population parameter that is generally unknown. Previously published research and statistical theory help guide the researcher. But with an audit, the auditor selects samples from computerized accounting systems. Such accounting systems can automatically summarize account totals and breakdowns; therefore key population parameters are generally known. In the exhibit below, the precision is 4%. Once the sample is drawn, one can use that sample to estimate a known population value from computerized reports.

Table 2.4 Trade-off between increase in efficiency and sample size

Detail	Detail Size	Sample Size n	Sample + Detail	Efficiency
650	349	436	785	0.86
850	284	479	763	0.84
1050	256	488	744	0.82

Suppose the precision in designing an audit sample is set at 4%. A validity check would be to determine if an account total sample estimate does indeed fall within the 4% of the audit population true value. Below is a design based on a detail of $850, 5 strata, and a precision (margin of error) equaling 4% with the resulting strata boundaries and sample size of 763 (Fig. 2.1).

When button #1 is selected, the account breakdown is displayed. In addition to defining stratification boundaries, sample size is displayed on last column (Fig. 2.2):

The last column is sample size by strata, and the MS Windows version can generate an Excel file documenting the random sample.

To generate the sample and validate the sample, select:

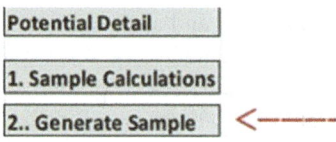

Then a new button will pop up to do the validation procedures. In the Windows version, a spreadsheet of the sample will also be generated, which involves a Mean per Unit Projection.

Below is a summary of population and sample.

The last line in the exhibit indicates that the validity check has been passed. To see how mean per unit estimate is calculated, *copy and paste the bottom screen into*

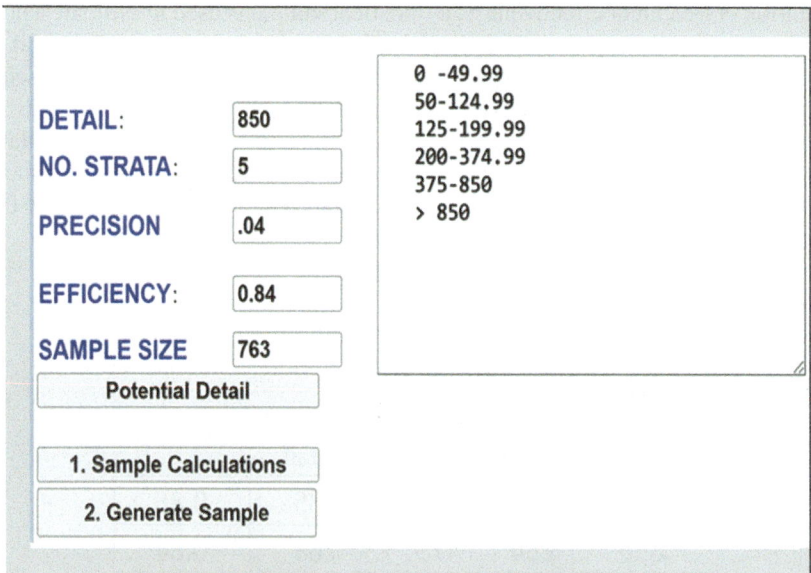

Fig. 2.1 Inputs to generate a sample with a detail stratum cutoff of 850

Population Specs:				
Freq. (N) Mean		Std. Dev.	Total Acct	Sample (n)
2872	25.11	14.98	72122	31
2328	81.43	50.22	189564	89
2799	156.05	41.99	436775	94
1674	248.25	90.72	415567	127
523	413.09	303.71	216065	138
284	2226.98	2228.09	632464	284

Fig. 2.2 Account stratification and sample size by strata

Population:		
N	Mean	Total $
2872	25.11	72122
2328	81.43	189564.2
2799	156.05	436774.8
1674	248.25	415567.5
523	413.09	216045.6
284	2226.98	632463.7

Sample:			
n	Mean	SD.	Total $
31	27.11	13.73	840
89	81.73	55.23	7274
94	161.92	20.94	15220
127	245.33	108.67	31157
138	419.38	310.69	57875
284	2226.98	2228.09	632464

Observed sample precision under 0.04

no need to resample←

Fig. 2.3 Audit population and sample summary indicating sample precision estimate meets the targeted value

an Excel spreadsheet in order to conduct the appropriate analysis provided below (Fig. 2.3).

The message at the bottom of the screen is:

Observed sample precision under 0.04 no need to resample.

Below is a discussion of how to calculate the mean per unit estimate. When going through this exercise you will have different numbers because you are working with a different random sample

2.4 Mean-Per-Unit Projection

After you cut and paste the above screen on to an Excel file then manipulate the columns as displayed in Table 2.5. You add the last column which is the mean of the sample multiplied by the total population frequency which is the shaded columns. The precision used in the sample size calculation was 4%. Does the selected sample using the mean-per-unit sample projection actually estimate an account total within the 4% target? The mean-per-unit projection method consists of multiplying the mean of each sample stratum by the total number of items (N − freq.) in the total population base. The mean-per-unit method sample estimate is a total of $1,339,630 with the true total dollar account total being $1,333,074 below the 4% threshold.

The last column to the right is mean per unit estimate of account totals by strata with the grand total indicating there is no need to draw a new sample. In fact, the difference between the sample estimate and the actual account total (audit population) is less than <1%.

On the validity check web screen, a popup link, *"check sample stats"* links to another web page that compares the sample estimate of error rate with the population's actual error rate. Also included on the screen is a breakdown of the sample into individual healthcare provider locations.

For this case study, a thorough full audit of the audit population was conducted. So, the parameter is known. The purpose of this sample was to determine which healthcare providers were not meeting claims reimbursement standards. As can be seen, the distribution was uneven (Fig. 2.4).

The audit population actual error rate is 10.2%, and the sample estimate is 10.5%. Third party administrators (TPA) claims processors would be very concerned if the error rate was ≥ 3%. Less than 3% is a reasonable target with a higher value to require reimbursements. In employer-based health plans, time lags in the communication of company human resource employee eligibility updates and varying timing of medical office submission procedures lead to some data inconsistency.

Table 2.5 Copy and pasted Excel file of mean per unit sample estimate of account total

Pasted From the Screen							
Population:			Sample:				
N Pop. Freq.	Pop. Mean	Total $	n	Sample Mean	SD.	Total $	Sample Mean X Pop. Freq.
2872	$25.11	$72,122	31	$24.20	$10	750	$69,502
2328	81.43	$189,564	89	$78.40	$57	6978	$182,515
2799	156.05	$436,775	94	$162.02	$21	15230	$453,494
1674	248.25	$415,567	127	$254.69	$68	32345	$426,351
523	413.09	$216,046	138	$397.26	$319	54822	$207,767
284	2226.98	$632,464	284	$2,226.98	$2,228	632464	$632,464

$1,330,074   Mean per Unit Total Estimate -> $1,339,630

Population Error Rate:

Plan	total Amt.	Error Amt.	Rate
Acme	$1,962,538	$200,285	0.102

Actual Error Rate= .102

Sample Results:

Loc.	Sum	Error	Rate
HMO1	$575,936	$62,146	0.108
HMO2	$120,018	$8,384	0.07
MD	$52,534	$588	0.011

Sample Estimated Error: 0.105

Fig. 2.4 Comparison of actual audit error rate and observed rate broken down by HMO plan location

For publicly funded plans like Medicare, there are a multitude of regulations that a healthcare provider's billing system may miss. A knowledgeable accountant armed with a streamlined tool in conducting an audit as in this case study has the best path for self-correction and self-protection.

A health plan's claims administrator provides feedback to both the plan member and healthcare provider. The plan member is sent an "explanation of benefits" statement that details all the claims paid and denied and what is the patient's responsibility. A similar report is sent to the healthcare provider. Being equipped with this feedback is one way for the auditor to have a path to enhancing an organization's knowledge of its administrative claims processing needs.

Quality management is the ability to anticipate. An auditor would be chagrined to realize there is a serious cashflow problem only after payroll cash needs cannot be met. It would be better to periodically obtain snapshots of current and recent past activity that paves the way for solving problems when they are relatively small. The same is true for assessing health plan administration and performance. At some time, an audit by an external agency may be mandated, and that is not the time to start preparations. Large corporations have resources to afford experts and sophisticated software in-house. It is the mission of Auditmetrics and this book to provide a means of professional development and efficient administration that evens out the analytic playing field for middle- and small-sized firms.

Part II
The Statistical Audit

Chapter 3
Statistical Audit Design

AICPA Standards—According to AICPA Statement on Auditing Standards (SAS) No. 39, the essential feature of statistical sampling is:

1. Sample items should have a known probability of selection, for example, random selection.
2. Results should be evaluated mathematically, in accordance with probability theory. Also covered are the mathematical techniques to assure efficient statistical estimates.

If just one of these requirements is met, that does not mean that the audit is statistical. For example, auditors, taxpayers, and others will sometimes use a random number method to select the sample and state they are using statistical audit methods. They may even enhance sampling efficiency by stratifying the random selection process. However, this is not statistical sampling if no attempt is made to evaluate sample findings mathematically (requirement number 2 above). The general statistical discussion that follows summarizes the statistical procedures specifically designed to conform to SAS No. 39. The fundamental principle of statistical sampling is that, although we do not have the resources to examine all transactions in a taxpayer's book of transactions, we can get a fair and unbiased estimate from a smaller manageable subset (random sample) if we follow both SAS No. 39.

There are two basic types of data sampling techniques, either variable sampling or attribute sampling. When data points are measurements on a quantitative numerical scale, they are variable data, e.g., weight and length, and, in the statistical audit, dollars. Variable sampling is standard for sales and use tax sampling or sampling in any situation where one wants to measure specific dollar quantities, such as total dollars in error or in compliance with regulations or other standards. Variable sampling is the standard for many states in the audit of sales and use transactions. The Internal Revenue Service (IRS) has established statistical guidelines when projecting revenues and/or expenses from a statistical sample. Variable sampling

J. Boffa, *AI Assisted Business Analytics*,
https://doi.org/10.1007/978-3-031-40821-2_3

techniques are mostly mathematical formula driven. The fundamental goal of variable sampling is to answer the question of quantity or "how much?"

Attribute sampling transaction data are classified categorically. For example, data tracking whether accounts receivable items are past due could be categorized as "yes" or "no." Attribute sampling is integral to opinion polling and market research where the pollster seeks the characteristics of targeted subsets of the population. In those environments, the data stratifications are situation dependent. For example, a pollster or market researcher may be interested in demographic breakdowns of potential customers such as socioeconomic status and gender. There are multiple variations of attributes, but to the accountant, they should ultimately relate to that all-important variable, dollars.

Attribute sampling is classification dependent. For example, an accountant may want to examine customer base in terms of sex, age category, or other sociodemographic classifications. These are attributes that are mutually exclusive, and the purpose of the audit would be to answer the question "how many?" are there in each category. There are several types of attribute questions that may be very useful for internal auditing purposes.

Examples of typical attribute sampling tests are:

- 20 of 100 accounts receivable invoices were past due.
- 10 of 40 inventory invoices greater than $1000 contained a signature.
- 19 of 20 fixed assets purchases had a supporting authorization document.
- 2 of 11 supplier invoices indicated the early payment discount was not taken.
- 13 of 211 journal entries were posted to the wrong account.

The results of an attribute sampling test, such as those above, are then compared to a criterion previously established. If the test results are worse than the standard, then the test has failed, and accounts should be carefully examined for possible remedies. For example, if the acceptable proportion of past due accounts receivable invoices is 3% and the tested rate is 20%, it may be necessary to impose additional controls, retrain staff, and/or alter invoice management procedures to reduce the number of past due invoices.

There are times when an auditor can exploit the benefits of both methods of sampling. The auditor may be interested in revenue and cost estimates (variables) by different segments (attributes) of the business or class of transaction. As mentioned previously, variable sampling procedures are easier to design and implement because the methodology is largely formula driven. The quantitative nature of dollars also provides more information about each unit of observation and therefore, in statistical terms, a more powerful estimate. An attribute sample is a collection of individual observations based on a common classification. Therefore, each data point is part of a collection of observations, each indistinguishable therefore lacking a certain amount of specificity rendering projections with less statistical power.

The Hybrid Approach—The design of the statistical audit sample methodology combines the characteristics of both variable and attribute sampling techniques. This combined approach is the methodology many states use as their standard for sales and use tax audits. In those audits, the auditor is interested in "how many" transactions in a taxpayer's account book are in error, such as having a tax not being paid or having paid a tax in error and deserving a credit. But the auditor is not only interested in "how many" transactions are in error, but also what is the dollar volume owed to the state, or "how much" owed in dollar value?

It would be impractical to examine all transactions, but a well-chosen representative subset or random sample can be used to project the total dollar amount. The statistical formulas discussed in this book allow the auditor to minimize the amount of statistical error when performing such projections. The hybrid process is a two-stage estimation process. First the auditor sets up a variable sampling technique to obtain a representative random sample of an account based on dollar value. Once the dollar based random sample is drawn, the auditor goes through a series of steps to determine which transactions pass or fail the audit's criteria.

Because accounts are commonly processed in all businesses electronically through standard software such as QuickBooks®, key economic characteristics of the total book of transactions are known. With account parameters known, it is possible to statistically test whether a sample statistic is valid and not an outlier. For example, if a sample estimate of total book value varies greatly from the known book value derived from accounting software, then the sample validity can be called into question. If that particular sample is termed a "statistical outlier," then a new sample should be drawn. There is a myriad of variables that can be validated in this manner including revenue, expenses, tax credits, etc.

Once this sample validation step has been completed to the auditor's satisfaction, the auditor then examines each transaction in the sample to determine if it is in error or if it satisfies established criteria. In the sales tax audit, the auditor is making what is called a dichotomous decision: either a transaction was properly paid or is in error. It is a simple yes/no decision and essentially sets up a two-category attribute which can be used to determine an error rate for the transactions sampled. Auditmetrics software and templates can be used to statistically estimate the total dollar volume in error.

The sampling exercise for sales and use tax audits starts with a variable sample and then morphs into an attribute sample application. Many states and national firms through the Multistate Tax Commission endorse these procedures for sales and use tax audits. In fact, the Multistate Tax Commission offers its own statistical audit courses and learning simulation programs.

The examples discussed are concerned with the primary question about dollar volume. However, dollars do not operate in a vacuum. There are always subsidiary personnel, customer, and organizational issues that need to be explored. Take the example of a medical claims adjudication case. In addition to determining the dollar error rate, there should also be an examination of attributes about specific staff, claim types, medical office structure, and health plan design characteristics. The auditor starts with an assessment of dollar volume slippage in the system, but from that point, there has to be determination which attributes or categories of transactions should be followed up.

Defining the Audit Population—The audit population is the total book of account transactions from which the sample is to be drawn. One of the first and most important decisions an auditor must make is to determine which transactions should be included in the audit population and which transaction should be excluded. For example, clients or MIS personnel may provide an electronic copy of an entire account file. However, many transactions in the file may not have implications for the specific audit. For example, if the audit is for a sales tax audit, not all transactions are subject to sales tax. Examination of transaction not subject to sales tax would be a waste of time for both the auditor who must examine the transaction and the personnel who have to pull those transactions. Therefore, it is up to the auditor to eliminate from the audit population as many irrelevant transactions as possible.

It is important that the auditor understands the account's list of transactions and is able to make an informed decision as to which transactions should be included in the audit population and which can be removed. The potential effect of including accounts with no implications for the purpose of the audit reduces sampling efficiency. The potential effect of excluding accounts that do have specific implications is to reduce the validity of the audit results.

Stratification—Stratification is the process of dividing the population of transaction into segments (strata) based on a certain characteristic. In variable sampling, one would stratify the population based on the dollar amount of the transaction.

One would stratify on the dollar amount of the transaction to accomplish the following:

1. Gaining sampling efficiency. A stratified random sample will yield more precise results than an unrestricted random sample of the same size.
2. Offsets the effect of extreme values (skewed distributions). To minimize the effect of large invoices on sample estimates, it is better for the auditor look at 100% of the invoices in the largest dollar stratum. This stratum is called the "detail stratum."

Stratifying the audit population is accomplished by using the dollar value of the transaction as the basis of stratification. Strata boundaries are determined by breaking down dollars into a collection of mutually exclusive categories. The recommended number of strata can vary from 3 to 10 which is also the Auditmetrics AI standard. There are several statistical techniques available for determining the most efficient basis for stratifying a population. The first question is how many strata and what should be the cutoff where 100% of the transactions are reviewed? There are some rules of thumb to reasonably determine the number of strata. However, the first step is to define the detail strata, where 100% of the transactions are reviewed. By eliminating the largest transactions from sampling, we reduce the variability (standard error) of the remaining transactions from which a sample will be drawn. This greatly enhances statistical efficiency. There are rules of thumb to determine what size of transaction should be the cutoff for the detail stratum. For example, one rule of thumb is to select all large transactions that account for 25% to 35% of the total book value. The detail stratum isolates those high dollar transactions that have the greatest economic impact.

This rule of thumb is useful in most cases. However, if you have access to and can use statistical software like SPSS, SAS, or STATA, then you can use the guidelines implemented in the Auditmetrics software. It uses a distribution-based criterion. It starts with examining transaction sizes that range from the 90th percentile to the 99.5th percentile.

The table below shows an audit population with a total of 60,916 records. The second column represents the value of the 90th to 99.5th percentile ranging from $901 for the 90th percentile to $26,145 for the 99.5th percentile. When deciding where one should establish the detail cutoff, a certain amount of judgment is required, and past experience is very helpful. It appears from this exhibit that a reasonable starting point would be somewhere between the 99th to the 99.5th percentile. The number of records would be between 609 and 304. That would be between $11,610 and $26,145 for a reasonable detail cutoff. The final decision for the population_test.xls was to set $16,000 as the detail cutoff. The Auditmetrics software goes through an iteration process and based on the criteria above suggests a reasonable detail cutoff for the auditor (Table 3.1).

Auditmetrics provides the auditor with these percentile ranks from which an AI-guided detail cutoff is generated. Auditmetrics will suggest a detail cutoff, but the auditor can fine-tune.

Once the detail stratum is established, there are several methods to determine the number of the remaining strata resulting in the most efficient sample in estimating audit population parameters. If you haven't a chance to complete the case studies, this is a good time to do so. One of Auditmetrics inputs is the number of strata. This is an opportunity to vary different number of strata and determine its impact on the efficiency factor.

Strata Boundaries and Sample Allocation—Of the various methods available to determine strata boundaries, many states have opted to use the cumulative of the square root of the frequency method. This method allows for greater efficiency in the sampling process compared with setting the boundaries using judgment only. Using judgment in setting up stratum boundaries may not be the most efficient. A problem with cumulative of the square root method is that there is no rule of thumb as a quick shortcut. The process can be tedious. With a working knowledge of Excel, one can systematically determine strata boundaries step. However, Auditmetrics AI

Table 3.1 Auditmetrics-generated percentile ranking of audit population

Percentile	Value	Cumulative	Strata Size
90th	$901	54,825	6,091
95th	$1,890	57,871	3,045
97th	$3,282	59,089	1,827
99th	$11,610	60,307	609
99.5th	$26,145	60,612	304

software automates this process, and if one would like a more detailed discussion, there are plenty of web sites that explain this method.

Once the strata boundaries are set, then the next step is to calculate a total sample size and then distribute that total into the strata. The method of sample allocation among the strata providing the best statistical efficiency is the *Neyman allocation*. The methodology takes into account both transaction count and inherent variability **S** (standard deviation) for each stratum. Steps in the process:

1. Calculate product of the frequency **(N)** in each stratum times the standard deviation for that stratum **(s)**.
2. Sum the products **(N × s)** for the all strata.
3. Calculate percentage for each stratum **(s X N) / Σ(s X N)** that is used to allocate total sample size **(n)** to each stratum.
4. This result is the number of items to be sampled for each of the strata.

Neyman allocation relies on the fact that the number of sampled units from each stratum is directly proportional to the frequency of that stratum and to the standard deviation of that stratum. For a constant sample size, Neyman allocation provides the smallest sampling error.

Table 3.2 outlines the Neyman allocation process. In this table is a two-step process; first calculate the total sample size, and then allocate that sample among the strata. Stratification based on a variable such as dollars means that as one proceeds up the strata ladder, the standard deviation increases in value which results in a rise in variability or statistical uncertainty. Allocating more of the sample n to the higher strata results in the minimization of the standard error of sample estimates.

The methodology discussed in Table 3.2 is a two-stage process. Auditmetrics has developed an algorithm that calculates the sample n for each stratum in one step that is more efficient than the two-step process.

Table 3.2 Neyman allocation

Strata:	Sample Size (n)	Mean	Std. Dev. (S)	Total Frequency (N)	N X S	Percent
0 -49.99	75*	$19.02	$14	10,101	$142,929	11%
50-174.99	132	$91.67	$43	5,776	$250,563	19%
175-399.99	141	$258.69	$95	2,813	$268,079	20%
400-824.99	157	$548.66	$185	1,614	$298,816	23%
825-1600	191	$1,106.04	$409	888	$363,521	27%
> 1600	463	$1,880.24	$0	463		
Excluding Detail	695			21,655	$1,323,907	100%
Total Including Detail	1,158					

* 11% X 695 = 75

$$n_i = \left(N_i SD_i\right)\left(\Sigma\left(N_i SD_i\right)/\left(A/U_R\right)^2 + \Sigma\left(N_i SD_i^2\right)\right)$$

Where:

n_i = sample size per stratum i
N_i = population size of stratum i
SD_i = standard deviation of stratum i
A = acceptable precision
U_R = reliability factor or Confidence level

Auditmetrics Statistical Inputs—No matter which method is used in determining sample size, it is important to emphasize that sample size is based on assumptions derived from statistical theory, particularly the expected precision, statistical confidence level, and assumptions about the distribution of the transactions in the audit population. Below are the statistical criteria of strata sample size used by Auditmetrics which is consistent with AICPA and IRS standards:

1. Confidence level: 95% confidence level
2. Required precision: 3–10% precision (margin of error) default 3%
3. Minimum sample size (30) per stratum defined by the central limit theorem
4. The number of strata 3–10 default at 6

Precision or as expressed in opinion polling margin of error is the primary driver of sample size. Three percent is considered the gold standard that would result in the largest sample size. But there are times when an auditor may want to conserve resources and draw a small sample to get some sense of the account's distribution.

The confidence level is an index of certainty. Generally expressed as a percentage or dollar range, confidence level refers to the probability that a true audit population dollar value will fall within a specified range. True population value is the dollar value that can be determined by 100% examination of the account. For example, if an audit population has an average of $100, and sample estimate has a 95% confidence interval (CI) around that, audit population's actual average is ± $3 or between $97 and $103. This would indicate that 95 random sample estimates out of 100 would be in that range that contains the true value of $100. That would also mean 5% of sample estimates would be outside of that range.

Precision and confidence interval are linked. For example, suppose we know a total account book of transactions has a mean of $100. If the goal is or a sample precision to be 33 of the true book average (audit population). The 95% confidence interval range would in turn be 6 (97–$103). A precision of 3% would require a larger sample size than a precision of 5% or 10%. Both confidence interval and precision are inputs needed to determine sample size.

Precision and confidence levels are generally used together in describing a sample design. A more precise way of expressing this relationship between 3% precision and 95% confidence is that one is 95% confident that sample the estimate would be within 3% of the true population value.

It is recommended that the minimum sample size in a stratum be 30. In cases where the stratum calculation is less than 30 transactions, one should increase that stratum's sample size to 30. A sample of 30 is usually considered the smallest sample size where the central limit theorem based on the normal- or bell-shaped curve can be safely used in calculating precision and confidence interval. The central limit theorem says that the sampling distribution of a sample estimate will always be normally distributed, as long as the sample size is large enough. A more detailed discussion of the central limit theorem is discussed in Appendix I.

One important observation to note from statistical theory is that there is no direct relationship between the audit population or account size and sample size. This is because the amount of variation or standard deviation of the dollar spread has a greater impact on sample size than simply the number of transactions to be sampled. A population of a million records with a large spread of dollar values or standard deviation is going to require a larger sample than a population of 10 million records with a very small dollar spread. The more diverse account requires a larger sample size to maintain precision and confidence interval standards as compared with one less diverse.

Chapter 4
The Sales Tax Audit

For this chapter it is expected the reader has taken the time to review the online video and had gone through an exercise of sampling using the practice audit population file available online. If you haven't or still feel unsure, then take the time to download "gettingstarted.pdf." For those who only have an Apple computer, review Case Study 2.

1. Case Study 1:
       Auditmetricsai.com
2. Case Study 2:
       Copy and paste on your browser:
https://healthe-link.com/audit/eaudit.aspx for the Auditmetrics link

It would also help those who have access to the Windows Version of the software to also download auditresult.xlsx:

https://auditmetricsai.com/filelibrary/auditresult.xlsx

It has an example of audit results that will be covered in this chapter. It documents statistical calculations discussed in this chapter.

The Excel file documents:

1. Stratification and parameters of audit population including efficiency factor documentation
2. Generated random sample statistics including mean, standard deviation frequency, and total dollars
3. Ninety-five percent confidence interval around each stratum to test whether each sample strata mean is statistically different from population mean
4. Audit results including ratio estimate of projected tax dollars owed including standard error estimate

Included in the work area of this file is an exhibit of different sampling strategies as they relate to standard error.

J. Boffa, *AI Assisted Business Analytics*,
https://doi.org/10.1007/978-3-031-40821-2_4

Efficiency Factor This is a measure of statistical efficiency by comparing a simple non- stratified random sample standard error with that of a stratified random sample. In the context of the statistical audit, the detail stratum is not sampled but reviewed at 100%. Standard error is a measure that estimates the efficiency, accuracy, and consistency of a sample estimate. In other words, it measures how precisely a sample statistic represents a population parameter.

For example, suppose an account with 10,000 transactions has a mean of $1000 and a standard deviation of $80 ($\sigma$) or variance of $6400 ($\sigma^2$). These measures represent the parameters of the account. A random sample of size 50 (n) is selected with a mean of $1050. Does this sample mean represent a reasonable estimate of the actual audit population account mean? It is the standard error that is used to make this decision:

$$SE = \sqrt{\sigma^2 / n}$$
$$SE = \sqrt{6400 / 50}$$
$$SE = \$11.31$$

Confidence interval plays a role in determining if one can conclude the sample mean of $1025 is a reasonable estimate of the population parameter $1000. Standard error will provide a range of statistical uncertainty around the population mean.

One standard error range = 989......$1000......$1011
Two Standard error range = 978......$1000....$1022

The key to statistical decision-making is to express the ranges above as a probability estimate. Sampling an account to calculate the mean can be done in an infinite number of times, and those means can form their own population called the "sampling distribution of the mean." Therefore, if a population has a mean μ, then the mean of the sampling distribution of the mean is also μ. The symbol μ_{mean} is usually used to refer to the mean of the sampling distribution. As with any distribution, there is a standard deviation of the sampling distribution which is given the name standard error. Of the standard error ranges listed above, the two standard error range is useful in making a decision about the reasonableness of the sample estimate of $1050. It involves a fundamental principle behind the mathematical properties of the random sampling process called "the central limit theorem."

The central limit theorem states that the sampling distribution of the mean will always be normally distributed, as long as the sample size is large enough, regardless whether the population has a normal, skewed, common with financial data, or any non-normal data. Auditmetrics AI software determines whether the normal curve applies with the sample statistics collected and makes adjustments.

The two standard error range indicates that 95% of sample means would be within the range of $977 and $1022. The 95% probability is based on the distribution of the normal curve. An alternative concluding statement is that 5% of sample means would be outside that range. Since the sample mean is $1050, the conclusion is it is outside the range of a reasonable estimate.

The more traditional statistical way of expressing this finding is that the sample mean is statistically different from the audit population mean of $1000. The conclusion is it is from a population where μ is greater than $1000. Can this be an error? Yes, 5% of random samples from the population with μ = $1000 would be greater than $1022 in this instance $1050. This is termed 5% alpha error. More detail about alpha error is discussed in Appendix I in the section discussing statistical decision-making.

The decision of the AI-assisted software is that $1050 is "not reasonable," and the software suggests to generate a new sample until the estimated mean is within the two standard error range. It is the size of standard error that is critical in guiding statistical decisions.

Efficiency factor is used as a guide to uncover the most efficient standard error. As the user varies sample design inputs, what is displayed on the screen is the efficiency factor. Table 4.1 displays the transition of standard error from a simple non-stratified random sample to that of a stratified random sample that excludes the detail stratum. The sample size is 695. Transactions greater than $1600 are reviewed at 100%.

The variance for a simple non-stratified sample is $141,623. The variance for a stratified random sample without a detail exclusion is $19,726. When the two variances are converted to standard error, the table indicates standard error is reduced by 62% indicating a smaller more efficient range in estimating the audit population mean. This improvement in efficiency is due to stratification alone. The variance for a stratified random sample that excludes transactions greater than $1600 is $11,439. As can be seen, the improvement in efficiency is now 72% which is displayed on the Auditmetrics screen.

The non-stratified sample descriptive statistics, mean, standard deviation, variance minimum, maximum, etc., are straightforward calculations using statistical software or Excel built-in functions. Calculation of the variance for stratified sampling process is more complex. It is more properly defined as stratification residual or error variance. It is identical in concept to the residual error around the regression

Table 4.1 The improvement of standard error from a non-stratified random sample to stratified sample with detail cutoff

	Variance (σ^2)	Standard. Error:	Reduction in Statndard Error
Non- Stratified Sample	$141,623	14.27	
Stratified Sample, no detail	$19,726	5.33	62%
Stratified Sample with detail	$11,439	4.06	72%

line previously discussed. The Auditmetrics algorithm focuses directly on calculating this residual variance.

The usual method of calculating stratification residual variance is to conduct an analysis of variance (ANOVA) procedure. It is a procedure commonly used in many scientific applications. For example, in a clinical trial, the stratification can be based on three different patient categories, those who take an experimental drug vs. those who take an established drug or those taking a placebo. ANOVA partitions variance among the stratified patient groups and the variance leftover is residual variance or inherent statistical error in the system. Using ANOVA procedures to tease out residual error would greatly slow down software performance. Auditmetrics V6.5 streamlined algorithm is valuable in handling large datasets. Professional Auditmetrics V6.5 has been tested sampling accounts with up to six million records.

As with all of the AI-assisted statistical calculations, statistical software STATA and SPSS were used to validate Auditmetrics calculations. As an example, ANOVA can be used to corroborate the residual variances in Table 4.1. Below is an ANOVA table using STATA statistical software to partition variance that corroborates the residual error variance of Table 4.1.

An astute statistician may point out that the sample size in Table 4.1 is held at a constant $n = 695$. Yes, the point is if you want the narrowest standard error range, then the sample size is 695. The role of the efficiency factor is to guide towards that efficient statistical estimation (Table 4.2).

But a question could be what sample size would be needed with for a non-stratified random sample with the same 3% precision and 95% confidence interval.

The formula for sample size for the same audit population but non-stratified is:

$$n = \left(Z_{05} \times \sigma / P \right)^2$$

where:
Z = 1.96 standard normal curve cutoff for 95% confidence interval
σ = standard deviation = $372
P (precision) = 3% × $202 (mean) = $6.06
n = (1.96 × $372/$6.06)²
n = $14,476

Table 4.2 Partition of variance using STATA's ANOVA

Source	SS	df	Variance	F	Prob > F
Between groups	1,293,400,000	4	323,352,304	28,267	0
Residual Error	242,365,481	21,187	**$11,439**		
Total	1,535,800,000	21,191	72473		

If sample size is a surrogate for cost, then the cost of this audit would be approximately 20 times higher than the stratified random sample with a detail stratum. The value of stratification is usually described as a more efficient method on two fronts, both cost and standard error.

It is the experience of the author that there are ranges of efficient factor that have different implications for action:

1. An acceptable efficiency should be ≥ .70.
2. Efficiency between 70 and 60 usually indicates a higher than usual skewed population. Testing different strata numbers and/or detail cutoffs may improve efficiency.
3. Between .50 and .60, a plotting or histogram of the data may reveal a pattern of extreme scores that may help in refining the specifications of the audit population.
4. In some instances, a factor in the 40% range may require the guidance of a statistician. The most common case is the combination of datasets are arbitrarily pulled together.

A very common cause of the low factor percent of item 4 is the bimodal distribution. It is the outcome of two processes with different distributions which are shown together in one set of data. It is also known as double-peaked distribution.

Figure 4.1 displays the data for a sales tax audit where several accounts were pulled together for the auditor's review. The bimodal display indicates that what was thrown together were independent revenue dynamics involving different product lines. The mean value is an arbitrary center point. The auditor was not able to obtain an efficiency factor in the 50% range. The suggestion was to divide the data into two separate audits. The efficiency factor for each was then in the mid 60% range.

Strata Confidence Interval Test   The audit population data for this exercise is population_test.xlsx. In this form it cannot be read into Auditmetrics V6.5. It has to be converted into a tab-separated text file by "save as (tab delimited) (.txt)." The purpose is to maximize data processing input output (I/O) performance. This is not relevant for the small business/learning version but is essential for the professional version when the file involves millions of transactions. More detail is available on the document "GettingStarted.pdf" available at:

https://auditmetricsai.com/FileLibrary/gettingstarted.pdf

Fig. 4.1 Bimodal distribution of account data

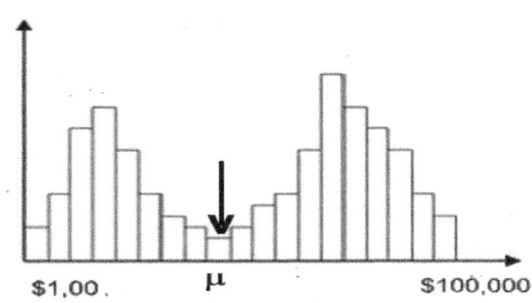

$1,00 .   μ $100,000

Our discussion so far is a sample with detail cutoff equals $1600, number strata equal 5, and precision is 3%. Figure 4.2 displays the Auditmetrics screen that indicates two validation tests have been passed. If either has failed, the auditor is to draw a new sample.

At the bottom of the lower screen is an indication that two validation tests have been passed. Validation #1 indicates that the total dollar volume projected from the sample is indeed within 3% of the true audit population total. The method used is the "mean per unit" projection. It is a total book estimate technique that multiplies the sample mean times the total number of items in the population. This is discussed in detail in Case Study 2.

Validation #2 is a 95% confidence interval around each stratum. The OKs indicate the success of the test. A confidence interval displays the probability that a parameter will fall between a pair of values around the mean. Confidence intervals measure the degree of uncertainty or certainty in a sampling method. Table 4.3 is the section of the AuditResult.xlsx Excel file that displays the strata specific confidence interval test. For the first stratum 0–49.99, the upper and lower bound for the 95% confidence interval around the sample mean is $15.55 and $22.97. This interval contains the mean of the population $\mu = \$19.02$. So, it passes the confidence interval test. The spreadsheet uses the Excel built-in function for the confidence interval. The formula for confidence interval is:

$$CI = \overline{x} \pm z \frac{s}{\sqrt{n}}$$

Auditmetrics - AI					
For Help: Info@auditmetrics.com			0 -49.99		
			50-174.99		
Detail	1600		175-399.99		
			400-824.99		
No. Strata	5		825-1600		
			> 1600		
Precision (Margin of Error)	.03				
Efficiency	0.72				
Total Sample	1158				

10101	19.02	192150.03
5776	91.67	529492.47
2813	258.69	727690.32
1614	548.66	885532.28
888	1106.04	982160.59
463	1880.24	870548.95

Potential Detail Cutoffs
○ Sample Size Excel File
⦿ Sample Validation Excel File

Sample Summary:
Validation Tests Listed Below

n	Mean	SD.	Total $		
67	18.75	14.92	1256	ok	ok
124	94.75	39.13	11749	ok	ok
138	258.23	102.98	35635	ok	ok
161	560.1	179.36	90176	ok	ok
205	1107.23	412.95	226981	ok	ok
463	1880.24	632.25	870549		

1. Sample Size Calculations

○ Select Another Audit Population?

Validation #1 - observed precision under .03, no need to resample

Validation #2 - strata specific test pased

Fig. 4.2 Sample specification indicating that all Validation tests have been passed

Table 4.3 Strata confidence interval test

Acme Inc.	Population	Sample			Strata Validity Test		
Strata	Pop. Mean	Sample Mean	Sample Std. Dev.	Sample Size	Lower 5% Alpha Bound	Upper 5% Alpha Bound	
0 -49.99	$19.02	$19.26	$15.48	67	15.55	$22.97	pass
50-174.99	$91.67	$91.07	$43.29	124	83.45	$98.69	pass
175-399.99	$258.69	$256.00	$93.33	138	240.43	$271.57	pass
400-824.99	$548.66	$544.32	$169.95	161	518.07	$570.57	pass
825-1600	$1,106.04	$1,080.24	$442.20	205	1,019.70	$1,140.77	pass
> 1600	$1,880.24	$1,880.24	$632.25	463			
Total (Excluding Detail)				695			
Total (Including Detail)				1,158			

CI Confidence interval
$\bar{x}$ Sample mean
Z Confidence level (1.96 for 95%)
S Sample standard deviation
$\frac{S}{\sqrt{n}}$ Sample standard error
For stratum 0–$49.99
CI = $19.26 ± 1.96 × ($15.48/$\sqrt{67}$)
CI = $19.26 ± $3.71
CI boundaries = $15.55 and $22.97

Ratio Estimation In audit sampling, the ratio estimation is a ratio or proportion of error in the sample which is used to estimate total audit error. This method applies the sample ratio to the entire population. Table 4.4 displays the section of the Excel template that calculates the total amount of transactions in error in that the business did not pay the required sales tax.

The auditor uncovered that the business owes a total of $126,007 based on ratio estimation plus $27,858 derived from the detail stratum for a total of $153,865.

The traditional way of expression the error ratio estimate is OE/BV × PBV where:

OE_i = Observed sample dollar amount in error of the ith stratum
BV_i = Sample book value of the ith stratun
PBV_i = Population book value of the audit population ith stratum
Ratio Error Estimate = Σ (OE_i/BV_i) × PBV_i = $126,007
Total Error Rate = $126.007/$3,317,026 = .038
For the 0–$49.99 stratum = ($54.20/$1290) × $192,150 = $8073

Ratio Estimate Standard Error The error rate above can be expressed as P = .038 where P is the probability of total dollars in error. This summary statistic involves a

Table 4.4 Ratio estimation of total amount of transactions in error where sales tax are owed

Acme Inc.	Population	Sample	Audit Results		
Strata	Pop.Total Value	Sample Total Value	Amount Error	Error Ratio	Pop. Est. Error Amt.
0 -49.99	$192,150	$1,290	$54.20	0.042	$8,070.30
50-174.99	$529,492	$11,293	$451.72	0.040	$21,179.70
175-399.99	$727,690	$35,328	$1,342.45	0.038	$27,652.23
400-824.99	$885,532	$87,635	$3,242.51	0.037	$32,764.69
825-1600	$982,161	$221,448	$8,193.58	0.037	$36,339.94
> 1600	$870,549	$870,549	$27,857.57	0.032	$27,857.57
Excluding Detail	$3,317,026	$356,995			$126,006.9
Including Detail	$4,187,575	$1,227,544			

statistical transition from variable data to attribute data. For example, two separate transaction dollar values can be:

Tx #1—Dollar amount for transaction #1 in the sample is $45.13
Tx #2—Dollar Amount for transaction #2 in the sample is $2.50

However, the auditor has made the decision that:

TX #2 is in error and taxes are due.
Tx #1 has sales tax properly paid.

The auditor has made what is called a dichotomous decision for each item in the sample; a transaction is either in error or not. So, the transactions now follow the binomial probability distribution. For the binomial conversion, there are only two outcomes:

Tx#1 = 0 not in error no taxes owed = 0
Tx#2 = 1 in error taxes owed $2.50
Etc.

There could be only two possible outcomes—yes or no—with the result being taxes owed or taxes not owed. This distribution is also called binomial probability distribution.

In Table 4.4 the total proportion or probability of the account total with taxes owed is 0.038 which is $126.007/$3,317,026. P now can be described as the probability of taxes owed and follows the normal approximation of the binomial probability distribution. The binomial can be approximated by the normal distribution with means $\mu = p$ and standard deviation $\sigma = \sqrt{n(1 - p)}$. This is valid when sample sizes are sufficiently large which is the case for the statistical audit.

Table 4.5 displays the post-audit statistical estimate is $126,007. Also included is a one standard error boundary. If one wants a 95% confidence interval, then multiply the standard error by two or more precisely 1.96. The detail error of $27,858 represents 100% of its transaction and not based on an estimate.

The standard error is based on the normal approximation of the binomial distribution:

P = Probability of the dollar amount in error = 0.038
(1 − P) = Variance of dollar amount in error = .962
n = Sample Size

$$\text{Standard Error} = \sqrt{\left(P \times (1 - P)\right)/n}$$

$$\text{S.E Dollars} = \sqrt{(038 \times .092)/n} \times 3,317,025 = \$24,053$$

Data and Methods The discussion has centered so far on the statistical audit process including its statistical theory underpinnings. The question at this point is: What is the primary source of cashflow information for the small business? All businesses have management information systems (MIS) responsible for user-friendly accounting software that tracks business income and expenses and organizes financial information. For small business one of the market leaders by far is QuickBooks®.

A valuable adjunct to a small business MIS system is to include a relational database management system (RDBMS). Appendix III discusses how to create a RDBMS using MS Access which is available with the professional version of Microsoft Office. For large businesses there are corporations such as ORACLE® that offer both a RDBMS and business MIS systems.

The building blocks for Auditmetrics data inputs can start with a MIS system to generate reports of sales, accounts payable, operating expenses, etc. A small business MIS system like QuickBooks® can generate account reports with an option that, in addition to a print version, generate an Excel version. As a demo, the

Table 4.5 Audit summary of taxes owed including ratio estimation standard error

Population Account Total=		$3,317,026		
	Std. Error($)=	$24,053	Sample Error Rate=	0.038
			Rate Include Detail=	0.037
	-1 Std. Error	$101,954		
Sample Error:	Mid-Point	$126,007	Detail Error:	$27,858
	+1 Std. Error	$150,060		
	Total Taxes Owed =		$126007 + $27,858 = $153,865	

Population_test.xlsx was loaded on a small business MIS system to generate a report of sales for the month of June.

The various headings of Table 4.6 are variable names relevant to this particular business. However, the last column is a required Auditmetrics input variable. If it has a different name on the MIS system, it should be changed to *"amount"* which will be the starting point to build an Auditmetrics input data matrix. First is to remove all of the Excel report embellishments such as heading, logo, and titles. The primary goal is to isolate the last column. Table 4.7 displays a sort of invoice number so that the data matrix floats to the top. The rows below the data matrix are superfluous and can be deleted.

Table 4.7 indicates that extraneous logo and report title were erased. The row at the top contains variable names with the data below. It was decided to sort on invoice number which allows a data matrix float up to the top. Sorting on amount may also embed a subtotal. There is no standard approach; the parsing of the report has to be decided on an individual basis.

Table 4.8 is the data matrix result of the sort. Also added are the required Auditmetrics variables. The variable absamt is created by the Excel built-in function = abs() and is sorted in ascending order. The variable DataSet is an identifier of the data source. Transaction_ID is a record count of the data files. Transaction_ID has value when the dataset is a merging of several data sources. For example, if you have two datasets of 1000 each and use MS Access to merge them into one file, then Transaction_ID 1 to 1000 is from the first dataset and 1001 to 2000 from the second dataset.

Table 4.6 Sales for ACME Co. broken down by Series and invoice number

	★ACME *Systems*	Sales - June			
8	PRODUCT	VENDOR_NUMBER	TX_CODE	INVOICE NUMBER	amount
9					
10	Series 6852				
11					
12	6852	3208	947	25546	556.19
13	6852	3208	807	25541	226.04
14	6852	3208	1002	25547	1065.8
15	6852	3208	1034	25545	1822.9
16	6852	3389	271	25402	-21.99
17	6852	3389	917	25401	439.75
18			Total		4088.68
19	Series 5860				
20					
21	5860	4800	507	18230	164.65
22	5860	4658	429	18078	746.26
23	5860	3272	483	18168	560.79
24	5860	3272	476	18156	-60.79
25	5860	3272	539	18184	75.99
26			Total		1486.9
	"	"	"	"	"

Table 4.7 Sorting report to isolate required data

	A	B	C	D	E
		VENDOR_		INVOICE_	
1	PRODUCT	NUMBER	TX_CODE	NUMBER	amount
2	5860	4658	429	18078	746.26
3	5860	3272	476	18156	-60.79
4	5860	3272	483	18168	560.79
5	5860	3272	539	18184	75.99
6	5860	4800	507	18230	164.65
7	6852	3389	917	25401	439.75
8	6852	3389	271	25402	-21.99
9	6852	3208	807	25541	226.04
10	6852	3208	1034	25545	1822.9
11	6852	3208	947	25546	556.19
12	6852	3208	1002	25547	1065.8
13	"	"	"	"	"
14					
15	Series 6852				
16					
17				Total	1622.02
18	Series 5860				
19					
20				Total	1622.02
21					

Table 4.8 Final sales data file with the required Auditmetrics variables

	A	B	C	D	E	F	G	H
		VENDOR_		INVOICE_				Transaction
1	PRODUCT	NUMBER	TX_CODE	NUMBER	amount	absamt	DataSet	_ID
2	5860	4658	429	18078	746.26	746.26	ACME	1
3	5860	3272	476	18156	-60.79	60.79	ACME	2
4	5860	3272	483	18168	560.79	560.79	ACME	3
5	5860	3272	539	18184	75.99	75.99	ACME	4
6	5860	4800	507	18230	164.65	164.65	ACME	5
7	6852	3389	917	25401	439.75	439.75	ACME	6
8	6852	3389	271	25402	-21.99	21.99	ACME	7
9	6852	3208	807	25541	226.04	226.04	ACME	8
10	6852	3208	1034	25545	1822.9	1822.86	ACME	9
11	6852	3208	947	25546	556.19	556.19	ACME	10
12	6852	3208	1002	25547	1065.8	1065.83	ACME	11
13	"	"	"	"	"	"	"	"
14								

The four variables at the right are those that are required. For more background on input data issues, Appendix II provides another example using the most commonly used small business MIS, QuickBooks®. Also, a more complete discussion of data issues are in "gettingStarted.pdf" is available on:

https://auditmetricsai.com/

Chapter 5
Forensic Accounting Using the Benford Formula

Benford's Law provides a data analysis method that can help alert CPAs to possible errors, potential fraud, manipulative biases, costly processing inefficiencies, or other irregularities. It is a relatively simple formula:

$$p(d) = \log_{10}\left(1/(1+d)\right)$$

Benford's Law gives the expected patterns of the digits, 0 to 9, in the numbers tabulated. Those digits, in unaltered data, will not occur in equal proportions; there is a large bias towards the lower digits, so much so that nearly one-half of all numbers are expected to start with the digits 1 or 2. The widespread applicability of Benford's Law is used to detect fraud, errors, and other anomalies. There are many examples on the Internet of authentic and accurate data that conforms to Benford's Law and the fraudulent and invented numbers that do not (Fig. 5.1).

The professional version of Auditmetrics AI has an added feature that when the detail cutoff stratum is determined, it can also perform a Benford's first-and-second digit test. It will also perform a goodness-of-fit chi-square test to determine whether the dataset being analyzed statistically conforms to the Benford formula.

There are some important considerations when performing investigations using Benford's Law. One is that the law applies only to naturally occurring data. Purchase amounts, payment amounts, stock prices, accounts payable, inventory prices, and customer refunds are all good examples of such data. It is important to avoid using financial data that are not natural. For example, the purchases at a discount store might not lend themselves to Benford analysis, because there may be limited price points per item. Similarly, values with upper and lower limits such as data sets of human heights, human weights and intellectual quotient (IQ) scores.

Some non-natural numbers when aggregated can be considered natural numbers, For, example, a restaurant's menu has set prices for specific items, not considered a list of natural numbers. However, when a total bill of all items is compiled into an

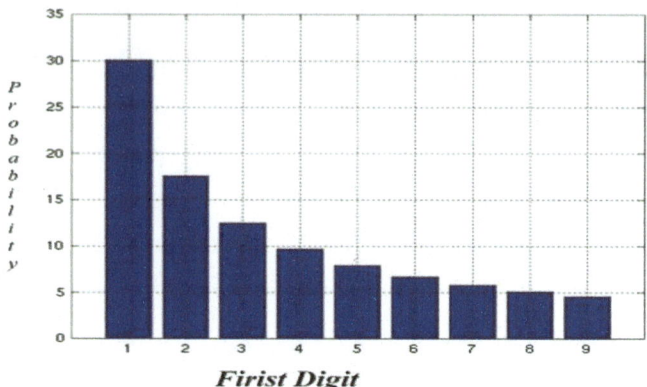

Fig. 5.1 First digit probability using Benford's formula

invoice, which is a sum of price times quantity, this can be considered a naturally generated set of numbers.

It is important to sample "fairly" when selecting a set of data for analysis. For example, limiting a sample of invoices to values between $100 and $999 defeats the use of Benford's formula because the data are limited to a narrow range. For small companies, using the complete data for multiple months or even a larger time frame is a better option than weekly or monthly reports. Benford performs best with large volumes of data.

Finally, as a technical matter, it is important to obtain a set of test data that is large enough to obtain useful statistical results. Auditmetrics AI uses chi-square test which compares expected observations based on Benford to what is actually observed, called goodness-of-fit chi square (Table 5.1).

Benford Example
Below is an Auditmetrics AI display of both a first and second digit Benford analysis of an account. Clearly the observed pattern deviates considerably from the Benford pattern:

As can be seen, the actual observed patterns for both first and second digit are considerably different from Benford's Law. Examination of the pattern indicates spikes at 4 and 6 for the first digit. There is no spiking for the second digit but a clear observable deviation from Benford formula around 5. Would this be an indicator of fraud?

If this is sales data and there is a common practice to offer price discounts resulting in high-volume purchases in the $40.00 to $60.00 range, then there is no indication of fraud. This would be an example of a non-natural dataset. However, this data is from an actual fraud case from an accounts payable account. The auditor decided to probe further and found there was unusual high variable activity by one specific individual with a specific supplier. It turned out the supplier was a sham where payments were sent.

Table 5.1 Conditions when the Benford Formula are likely and unlikely used

When Benford Analysis Is Likely Used	Examples
Sets of numbers that result from mathematical combination of numbers – Result comes from two distributions	*Accounts receivable (number sold * price),* *Accounts payable (number bought * price)*
Transaction-level data	Disbursements, sales, expenses
On large data sets – The more observations, the better	Full year's transactions ideal
Accounts that appear to conform is when the mean of a set of numbers is greater than the median and the skewness is positive	Most sets of accounting numbers fit this pattern
When Benford Analysis Is Not Likely Used	**Examples**
Data set is comprised of assigned numbers	Check numbers, invoice numbers, zip codes
Numbers that are influenced by human thought	Prices set at psychological thresholds of ATM withdrawals
Accounts with a large number of firm-specific numbers	An account specifically set up to record $100 or more refunds
Accounts with a built in minimum or maximum	Set of assets that must meet a threshold to be recorded
Where no transaction is recorded	Thefts, kickbacks, contract rigging

The reason it was not readily observable was that the fraud was conducted carefully, periodically, varying days and weeks and varying dollar amounts small enough not to be noticed over a long period of time, well hidden in the shadows. But the auditor collected data over a long period of time, and with a large volume of observations, the pattern of Fig. 5.2 emerged. Natural numbers under Benford follow logarithmic patterns work because fortunately fraudsters do not think in that same pattern and cannot generate logarithmic patterned numbers.

Unfortunately, the chi-square test with smaller datasets can lead to false negatives. Benford works best with very large datasets preferably with multi-thousands of data points. That is why it is only available in Professional Auditmetrics V6.5. smaller datasets have greater volatility. With smaller datasets the best way to determine Benford compliance is to look at the plot of the data. It is the spike points that may be the first indictor of potential fraud.

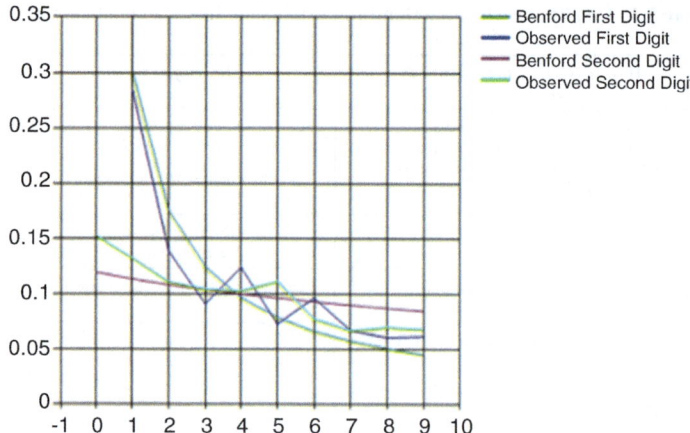

Fig. 5.2 Benford first and second digit graph with smooth lines following Benford's logarithm function

Part III
Forecasting Revenue and Expenses

Abstract Once a random sample has been used to monitor and improve a firm's cashflow, the next step is to use that same sample to forecast future trends. The fundamental techniques of the model building branch of statistics are regression and correlation. In addition to forecasting, regression and correlation techniques are also used to manage budgets and plan short-term investments for improving business operations.

Chapter 6
Financial Projections

Forecasting helps managers guide strategy and make informed decisions about critical business operations such as sales, expenses, revenue, and resource allocation. When done right, forecasting adds a competitive advantage and can be the difference between successful and unsuccessful outcomes. Many business managers complain that building forecasts with any degree of accuracy takes a lot of time. The time may be better spent selling rather than planning. But few financial institutions will put money in a business if it's unable to provide a set of thoughtful forecasts. More importantly, proper financial forecasts will help develop operational and staffing plans that will make a business more efficient.

Correlation and regression are powerful statistical tools for forecasting. The key is to draw random samples from relevant accounts, a topic which was covered in Part II. Auditmetrics software guides business managers in conducting statistical audits that are mathematically efficient, inexpensive, and easy to use. Just as in political polling, the starting point is to set a targeted margin of error, and the software then guides the user to the drawing of a fully validated and documented random sample.

Most business planners would prefer making revenue forecasts; it's easy to forget about expenses. Many will optimistically focus on reaching revenue goals and assume expenses can be adjusted to accommodate the incoming revenue. In this section is discussed using regression and correlation to forecast both revenue and expenses. It is our experience when preparing a business plan, start with expenses. Most small businesses fail because costs are inadequately defined and controlled. The best way to reconcile revenue and expense projections is by a series of reality checks for key ratios. Here are a few ratios that should help as a guide:

Gross Margin It's the ratio of total direct costs to total revenue during a given quarter or given year. This is one of the areas in which aggressive assumptions typically become too unrealistic. Beware of assumptions that make your gross margin increase from 10% to 50%.

J. Boffa, *AI Assisted Business Analytics*,
https://doi.org/10.1007/978-3-031-40821-2_6

Operating Profit Margin It's the ratio of total operating costs—direct costs and fixed cost overheard, excluding financing costs—to total revenue during a given time period. A growing business should show a positive movement with this ratio. As revenues grow, overhead costs should represent a small proportion of total costs. When developing a business plan, regression analysis is a valuable tool in developing projections of both revenue and expenses. Building an accurate set of growth projections for a startup will take time, but regression statistical tools will help paying attention to detail and expedite the process.

Scenario #1: Projecting Monthly Revenue This scenario starts as its base the statistical audit sample and convert for projecting monthly total revenue. The data in this section is from a multi-specialty medical group practice. The variable amount from audit file is what is to be forecasted. The random sample generated by Auditmetrics is a comma-separated variable (.csv) text file. It has the property that it can be read in directly by Excel. To proceed, the sample needs to be "save as" an Excel .xlsx file. Using standard Excel methods, the data then can be aggregated as seen in Table 6.1.

Below is an excerpt from 36 months of sales data:

The data now is in the form that can be used to create a regression model to use month count to predict monthly sales. Excel can be used to calculate the straight line values of intercept (a) and slope (b) that minimizes residual error variance $S^2y.x$. The error term in this model represents the variables that are unknown and not measured. Two other useful Excel functions are = Year(date) that generates the year and = Month(date) that generates month number 1–12.

Regression requires the Excel Professional version that includes the "Analysis Toolpak" and must be installed separately. It is not automatically installed with Excel. Installation is accomplished when a workbook is open, going to "file" and

Table 6.1 Excerpt from random sample of a 36-month sales data broken down by year and month count

Sales by Month

DataSet	Year	Month	MonthSales	MonthCount
MDPartners	1	Jan	20782.07	1
MDPartners	1	feb	15471.95	2
MDPartners	1	march	22623.79	3
MDPartners	1	april	21999.19	4
MDPartners	1	may	23199.81	5
MDPartners	1	june	23600.28	6
MDPartners	1	july	28824.02	7
MDPartners	1	aug	25719.38	8
MDPartners	1	sept	47264.88	9
MDPartners	1	oct	49389.22	10
MDPartners	1	nov	17723.06	11
MDPartners	1	dec	90301.02	12
MDPartners	2	Jan	31789.62	13

selecting "options" and then selecting "add ins" Analysis Toolpak. Once installed, click on "data" at the top menu, and "data analysis" will popup. Select regression, and for the Y variable, select "MonthSales" and the X variable MonthCount (Fig. 6.1):

Input Y range is the sales variable and X range is the month number. Since "Labels" is checked, the columns selected must include the top variable name at the top row. The default confidence interval is 95%. There is no need to change since it is the default for Auditmetrics sample selection. Also select "line fit plots." It is best to select "New Worksheet" which will contain the regression model summary and a line plot chart together.

Below is an excerpt of the most relevant portions of the results: what is Recipe, is it necessary?

R—0.95   a—$55,617
R^2—0.91   b—$14,598
n—36

In the model intercept = $ 55,617 and slope = $14,598
The prediction formula becomes Sales = $55,617 + $14,598 × Month Count

Fig. 6.1 Excel panels to implement regression analysis

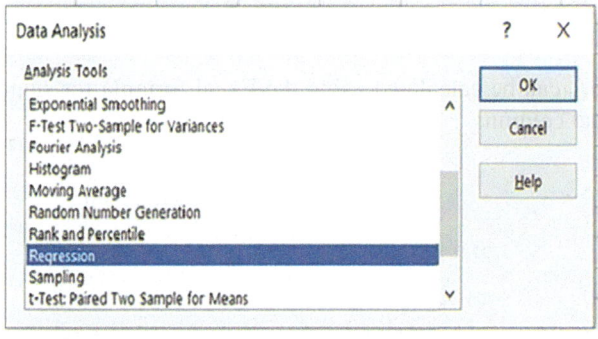

The correlation R of .95 and $R^2 = 91\%$ indicate a very good fit of the data to the linear model. Ninety-one percent of sales variability can be explained by the month count that varies from 1 to 36. This means the residual variance Syx is 91% smaller than the variance of just using the sales mean (Y_{mean}) as a predictor.

As a rule of thumb, a correlation greater than 0.75 is considered to be strong between two variables. Correlations between 0.45 and 0.75 are moderate, and those below 0.45 are considered weak. A weak correlation does not mean the business is failing. Its financials may be good; it just means it is not growing. The statistical model section of Appendix 1 goes into more detail discussing R and R^2 (Fig. 6.2).

The regression line is an estimate with statistical uncertainty as indicated by the data points around the straight line. Table 6.2 summarizes the data output of the model. The column is the observed monthly sales total or Y, predicted straight line Y′, and residual which is the observed minus the predicted. Residual is another name for the error term (**e**) in the prediction model:

$$Y = a + bX + e$$

Predicted Y′ is the straight line. If you plot this column using Excel, a straight line would be the result. The residual is an error term, the difference between the straight line and actual data point Y′-Y.

The residual (**e** = Y′ − Y) follows the normal distribution with a mean of 0 and standard deviation of:

$$S_{y.x} = \sqrt{\frac{\Sigma (e)^2}{n-2}}$$

$S_{x.y}$ can be calculated using the Excel formula for standard deviation of the residual column:

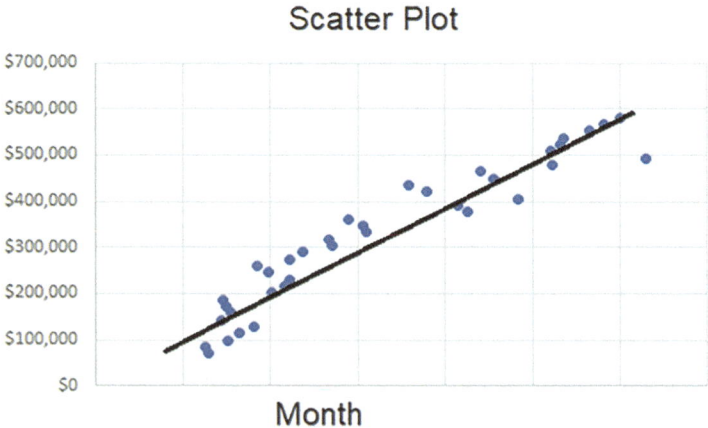

Fig. 6.2 Scatter plot of the relationship between month count and monthly sales

Table 6.2 Observed monthly sales, regression predicted monthly sales, and residuals

DataSet	Month	Y (Sales)	Predicted Y'	Residual(Y'-Y)
MDPartners	jan	$129,646	$70,216	-$59,430
MDPartners	feb	$126,227	$84,814	-$41,413
MDPartners	march	$151,966	$99,412	-$52,554
MDPartners	april	$165,168	$114,011	-$51,157
MDPartners	may	$182,977	$128,609	-$54,368
MDPartners	june	$145,694	$143,207	-$2,847
MDPartners	july	$156,814	$157,805	$991
MDPartners	aug	$150,373	$172,404	$22,031
MDPartners	sept	$146,155	$187,002	$40,847
MDPartners	oct	$202,621	$201,600	-$1,021
MDPartners	nov	$217,081	$216,198	-$883
MDPartners	dec	$222,458	$230,797	$8,339
MDPartners	jan	$198,338	$245,395	$47,057
"	"	"	"	"

$$S_{x.y} = \text{STDEV.S}(E2 : E37) = \$47,121$$

Scenario #2: Multiple Regression So far, the basic model is a bivariate linear model, a dependent variable with only one predictor variable. Though we have a very good fit, there is a problem with the model. Data is that of a wholesaler that supplies retail outlets. With this prediction model of the next month or quarter, it will always be higher than the previous month or quarter. But business activity does have seasonal fluctuations. The fourth quarter of the year and its holiday activity will always be higher than the following first quarter of the following year. The model as it exists does not allow for seasonal fluctuations.

We need a new model with two predictor variables that includes monthly and quarterly predictors using multiple regression. It is a statistical technique that uses several explanatory variables to predict the outcome of a dependent variable. In essence, multiple regression is an extension of ordinary least squares (OLS) or bivariate regression that involves more than one explanatory variable (Fig. 6.3).

The multiple regression model depicted is represented as a three-dimensional plane. Mathematically it is possible to expand the number of predictor variables into multiple dimensions which cannot be pictorially represented.

Setting Up Data to Adjust for Seasonal Fluctuations The first step is to determine whether if there is indeed seasonal fluctuation and then set up the data file for such adjustment. If there is evidence of fluctuations, the next step is to set up the data file with a quarterly input matrix. The first step is to use an excel "pivot table"

Simple and Multiple Linear Regression

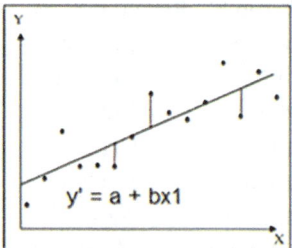

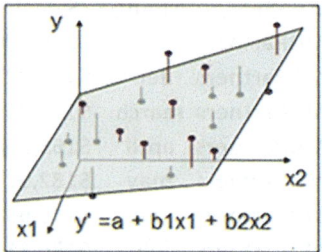

In simple linear regression the least-squares estimators minimize the sum of the squared errors from the estimated regression line.

In multiple linear regression the least-squares estimators minimize the sum of the squared errors from the estimated regression plane.

Fig. 6.3 Linear regression with one predictor variable and multiple regression with two predictor variables

to break down sales data by quarter. If quarterly fluctuations are detected, then set up regression input matrix with quarterly inputs.

Below is the data matrix that will be used to determine adjustment for quarterly fluctuations. Quarter column with 1–4 values uses excel function *ROUNDUP(MONTH(A1)/3,0) where A1 is the cell that has the date field in form Mo/Day/Year* (Table 6.3).

Setting Up Quarterly Input Variables Dummy variables are a special attribute variable in statistics and econometrics, particularly in regression analysis. It takes only the value 0 or 1. It indicates the absence or presence of some categorical effect that is used to shift the outcome. They can be thought of as numeric stand-ins for qualitative facts in a regression model sorting data into mutually exclusive categories (such as men and woman).

In the case of quarterly adjustments, we set up the data matrix with the following dummy variables:

- Quarter 1—1 if yes or 0 if no
- Quarter 2—1 if yes or 0 if no
- Quarter 3—1 if yes or 0 if no

You may notice the data matrix has dummies for three of the four quarters. It may seem logical to set up four variables Q1, Q2, Q3, Q4. The problem with this method is that the last dummy variable (Q4) is redundant. If the first three dummy variables are 0, then it is a given that Q4 has to be 1, so it is defined and is not random. The rule is:

1. The number of dummy variables necessary to represent a single attribute variable is equal to the number of levels (categories) in that variable minus one.

Table 6.3 Dataset with the addition of quarterly dummy variables

DataSet	Month	Quarter	Sales	Q1	Q2	Q3
MDPartners	jan	1	$129,646	1	0	0
MDPartners	feb	1	$126,227	1	0	0
MDPartners	march	1	$151,966	1	0	0
MDPartners	april	2	$165,168	0	1	0
MDPartners	may	2	$182,977	0	1	0
MDPartners	june	2	$145,694	0	1	0
MDPartners	july	3	$156,814	0	0	1
MDPartners	aug	3	$150,373	0	0	1
MDPartners	sept	3	$146,155	0	0	1
MDPartners	oct	4	$202,621	0	0	0
MDPartners	nov	4	$217,081	0	0	0
MDPartners	dec	4	$222,458	0	0	0
MDPartners	jan	1	$198,338	1	0	0

2. For a given attribute variable, none of the dummy variables constructed can be redundant. That is, a dummy variable cannot be a constant multiple, simple linear relation, or defined by another variable.

In the multiple regression formula, the predicted value for Quarter 4 is when Quarter 1 to Quarter 3 are all Zero.

Using Excel Function to Set Up Dummy Variables Rather than manually entering dummy values, one can program entries using Excel's "= if(.....)" function. Like every function and formula in Excel, IF is based on a specific syntax:

$$= IF(condition, value\ if\ true, value\ if\ false)$$

As shown above, the function has three parameters, the first two of which are compulsory.

Condition: This position must contain a condition—a comparison between two values— where one or both values can be cell references. The possible conditions are:

Equal (=)
Unequal (<>)
Less than (<)
Greater than (>)
Less than or equal to (<=)
Greater than or equal to (>=)

Below is the data matrix displaying the formula in the first row:

	A	B	C	D
1	Quarter	Q1	Q2	Q3
2	1	=if(a2=1,1,0)	=if(a2=2,1,0)	=if(a2=3,1,0)
3	''	''	''	''

Copy and paste it to the rest of the rows.

The full regression equation:
Total sales = $83,049 + (14,317 × Month)
+ (Q1 × −38,272)
+ (Q2 × −$8334)
+ (Q3 × −$42,340)

Comparison of a Quarter 1 estimate with Quarter 4:

Total Sales Month 36 (Quarter 4) = $83,049 + ($14,317 × 36) = $598,461
Total Sales Month 37 (Quarter 1) = $83,049 + ($14,317 × 37)
+ (1 × −$38,272) = $560,189

As expected, quarter one is less than Quarter 4.

Limitation of Regression in Forecasting Past data is used to predict future outcomes. Since our data has excellent R^2, prediction can be very reliable but up to a point. To understand what this means is that any statistical projection does have a certain amount of error. The regression line does have a 95% confidence interval band. Like any statistical estimate, the straight line has a band of uncertainty designated by a standard error. The 95% means that out of 100 samples predictions, 95 would be within the band that contains the population's true straight line, while 5 would be outside that band. Those prediction outside represents 5% alpha error as discussed in the section on statistical inference.

The confidence band around a mean is a straightforward constant interval around the mean. However, the confidence interval band around the prediction line is more complex. The actual calculation is a complexity that is covered in more advanced book on regression. The exhibit below depicts the 95% confidence interval around the regression line (Fig. 6.4).

Fig. 6.4 Ninety-Five Percent confidence interval around the regression line

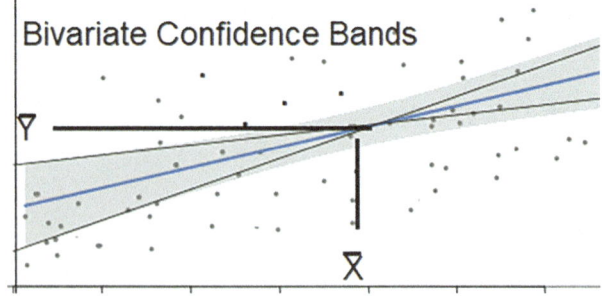

The band of confidence is the narrowest at the intersection of the means of the constituent variables. For example, for the simple bivariate regression line, the best prediction is at the intersection of the two means. As you go up the X scale, the bands expand, and predictions have steady wider confidence intervals. It is not that much of a problem with good fit of R^2 to project two or three quarters into the future, maybe a year if the database of the regression is of sufficient size. Projecting up to 12 months based on 6-month data will have wider and wider estimates.

As the new quarterly regression data is added to the database, it is best to drop and archive the earliest quarter. This keeps the data current. This is a way to start detecting any degradation of the model such as R^2 slipping which may be an indication of shifts in the market that may need further analysis.

Chapter 7
Planning Expenses and Investments

When beginning the process of forecasting, it may be easier to start with expenses, not revenues. It's much easier to forecast expenses than revenues since all successful businesses expend time and efforts to implement operational budgets to implement resource allocation and monitor performance. Regression has two statistical methods relevant in guiding the budgeting process:

1. *Linear regression*
2. *Curvilinear regression*

Variable budget, also called flexible budget, is a financial plan for estimated expenses based on linear regression monitoring productivity in relation to input costs. In other words, a variable budget uses the expenses produced in the current production of goods and services as a baseline to estimate how expenses relate to customer activity. Management uses variable budgets to predict best case and worst-case scenarios for an upcoming accounting period. This provides a "what if" look at future financial performance. The correlation coefficient is also used as a measure of how well expenses are controlled.

Variable budgets are appropriate for larger businesses especially those that have multiple locations. With small business all labor costs can be usually handled as a fixed cost. The classic method of budgetary control of monitoring the difference between budgeted and actual cost (variance) is an efficient method of control. Larger volume firms should take the time to categorize labor costs as they relate to productivity.

Below is an example of 12-month costs as they relate to production as measured by number contracts signed (Table 7.1).

An alternative can also be the dollar value of the contracts.

In this budget:

1. Contracts signed is the measure of productivity in delivering specified goods directly associated with the number of contracts signed.

Table 7.1 Listing of costs as they relate to productivity as measured by the number of signed contracts

Contracts Signed	Cost of Goods	Direct Labor	Other	Fixed Cost
3000	$14,500	$24,500	$21,000	$75,000
3800	$16,700	$31,400	$26,000	$75,000
5000	$20,000	$33,500	$31,500	$75,000
5800	$21,500	$38,700	$35,000	$75,000
4100	$14,050	$29,300	$24,100	$75,000
3100	$15,900	$26,600	$22,500	$75,000
3200	$20,100	$26,500	$22,500	$75,000
4800	$19,900	$24,100	$29,400	$75,000
4500	$17,900	$30,000	$28,600	$75,000
3700	$15,000	$29,100	$26,000	$75,000
5100	$22,500	$36,700	$32,000	$75,000
6000	$23,010	$39,500	$35,000	$75,000

2. Cost of goods is a variable cost that includes materials and supplies that are associated with the production of the goods.
3. Direct labor is a variable cost that is directly associated with the production and delivery of goods.
4. Other is a catch-all variable type labor category involving costs such as customer service and marketing outreach. These costs are not directly variable but can be termed semi-variable also known as a semi-fixed cost or a mixed cost. It is a cost composed of a mixture of both fixed and variable components. Costs are fixed for a set level of production, and they become variable after this production level is exceeded. In relation to productivity, they increase not in a straight line but in steps of increases.
5. Fixed costs are general overhead time costs such as rent, administrative salaries, insurance, property taxes, interest expenses, and potentially some utilities (Fig. 7.1).

This is a plot of all variable costs as they relate to productivity, contracts signed. Variable costs are zero if there are no contracts signed. Fixed costs are constant and do not vary with productivity. In order to control expenses, one must look at each individual cost type to determine how closely each is associated with output. Semi-variable is difficult to plot but can be statistically measured.

Correlation is used as the measure of the association of each cost with output.

X—Independent variable: Signed contracts

Y—Dependent variables:

- Cost of goods
- Direct labor
- Other expenses

Fig. 7.1 The plotting of fixed and variable costs in relation to productivity as measured by contracts signed

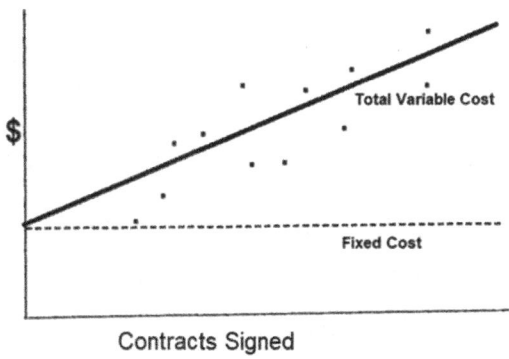

Contracts Signed

Correlation results R:
Cost of goods = 0.77
Direct labor = 0.82
Other = 0.98

With regression "what if" scenarios can be implemented. But there are some interesting findings in this table. It would be expected that variable costs (cost of goods) would have a better correlation with productivity than direct labor. A look at R^2 would give a better picture as a measure of cost control. The R^2 are as follows: cost of goods 59%, direct labor 67%, and other expenses 96%. Other expenses including marketing and customer service are efficient. In retrospect it seems logical that sales people and those who handle customer complaints have a direct impact on customer acceptance that leads to contract growth. Cost of goods are the materials used in the production process. However, they are associated with the production process. Is there inefficiency in the production process? It may be inefficient direct labor management, ineffective inventory control, or outmoded equipment. The first two will require some research in streamlining productivity of existing resources, but the latter may involve investment in new equipment. Such investments will involve long-term financing considerations and depreciation.

A variable budget can be developed using regression techniques.

The budget of Table 7.2 is based on expectations of most likely level of sales and expenses. This has advantages over the more traditional static single budget. It is in essence a budget with contingency plans for when and if customer demand either increases or decreases. For larger business it can be a roadmap for shifting resources. One department or entity may experience a downturn, a shift in resources to other areas where there is growth. A variable budget is more complicated and requires a solid understanding of a company's fixed and variable expenses. It allows for greater control over changes that may occur.

Nonlinear Functions One of the most common questions asked about regression is how well cost and revenue functions follow the straight line. Is the best model for forecasting revenue and controlling cost? Before going forth with analysis, it is always wise to plot the data and determine if the data does indeed seem to follow a

Table 7.2 Annual variable budget for fixed cost, cost of goods, direct labor, and other expenses

Aveage Signed Contracts/Month	2,917	4,167	5,000	6,250
Annual Contracts	35,000	50,000	60,000	75,000
Fixed	$75,000	$75,000	$75,000	$75,000
Cost of Goods	$83,442	$119,203	$143,043	$178,804
Direct Labor	$147,865	$211,236	$253,483	$316,854
Other Expenses	$34,276	$232,888	$279,465	$349,331
Total Budget	$340,583	$688,327	$750,991	$919,989
Budget: monthly regression formula X 12 for annual budget				

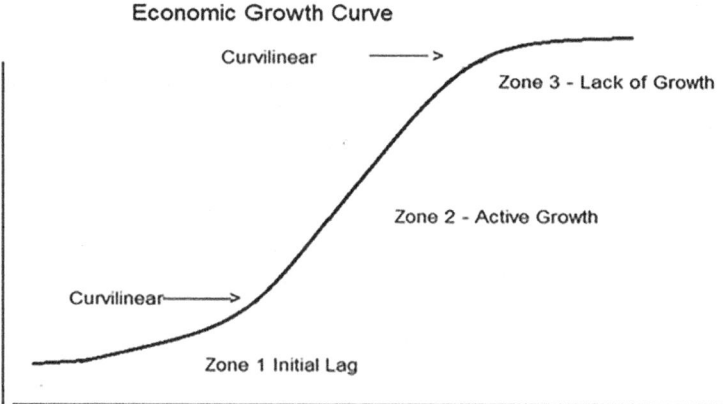

Fig. 7.2 Growth function of a business

linear function. Below is a plot of long-term growth of a business. There is an initial accelerated growth phase with an eventual slowing down and an eventual static final phase (Fig. 7.2).

Displayed is a growth curve for any business. There is an initial investment lag (zone 1) proceeding to steady accelerated growth (zone 2). Linear regression would be a good fit for zone 2. Zone 1 is not a good fit for standard linear regression, but there are other regression models that can be used.

Zone 1 is the critical initial phase in the growth curve where there is a startup lag in revenue while expenses start on day one. In this phase a careful examination of expenses and projected revenue is paramount. Below is an example of a log function of cashflow that is a good representation of Phase 1 growth. Figure 7.3 indicates cumulative cash inflow will pay back a $10,000 investment budgeted at $1000 per month in 11 months.

Fig. 7.3 Time in months
for cashflow to payback
initial $10,000 investment

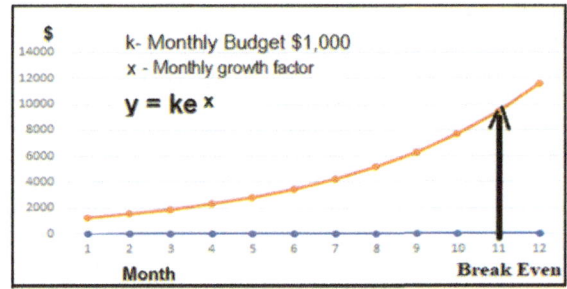

Table 7.3 Exponential cashflow growth function for short-term investments

Month	X	*Projected Cashflow	Observed Cashflow	**log Cashflow Projected	**Log Observed Cashflow
	Growth Factor inc. = .2	Monthly Growth k -$1000			
1	0.2	$1,221	$875	7.108	6.774
2	0.4	$1,492	$2,001	7.308	7.601
3	0.6	$1,822	$1,910	7.508	7.555
4	0.8	$2,226	$3,587	7.708	8.185
5	1	$2,718	$1,800	7.908	7.496
6	1.2	$3,320	$5,120	8.108	8.541
7	1.4	$4,055	$3,478	8.308	8.154
8	1.6	$4,953	$3,500	8.508	8.161
9	1.8	$6,050	$5,142	8.708	8.545
10	2	$7,389	$8,960	8.908	9.101
11	2.2	$9,025	$10,006	9.108	9.211
12	2.4	$11,023	$13,098	9.308	9.480

* Excel Function: =kEXP(x)
** Excel Function: =LN(Dollars)

The formula $Y = ke^x$ where e^x can be calculated using the Excel function $= EXP(x)$. The exponent X drives the incremental growth and **k**—is the projected monthly budget.

Linear regression can be used to determine the statistical reliability of the projected cashflow. This requires converting dollars into natural log values by using the Excel function = **ln(dollars).** The fitting of a straight line to the function of the curve will not yield efficient residual variances. However, linear regression can be used to describe exponential functions by linearizing the data set in question. That is the purpose of converting dollars to natural log values. The right two columns are dollars converted to the natural log.

This table's projected cashflow reached a payback after the eleventh month as was projected. The X in the formula is what drives the curvilinear logarithmic growth over time. A conservative starting point would be 0.2 and to increment 0.2 each month. This formula is similar to the exponential formula for compound interest growth. However, the formula of Table 7.3 is more conservative. The exponent in compound interest growth is $e^{(i+t)}$ where **t** is the compounding period.

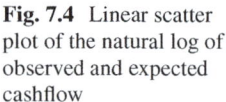

Fig. 7.4 Linear scatter plot of the natural log of observed and expected cashflow

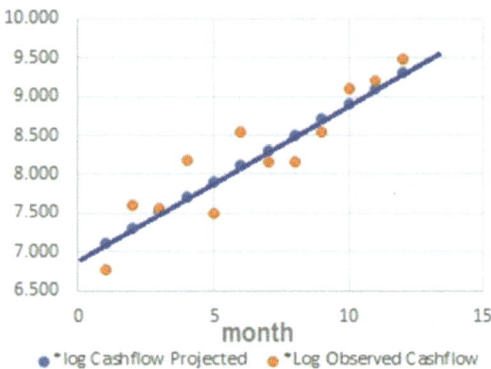

It should be noted that this business plan is intended for a cashflow horizon of a relatively short accounting period. It is intended for short-term business investments such as expanding advertising capabilities, hiring additional part time staff to quickly improve operational capabilities, or investing in new software and training to improve administrative efficiency. These are the types of business investments where the turnaround of cashflow would be 12–18 months. The increment of .2 can be adjusted to shape a shorter or longer turnaround.

This business plan assessment is not intended for long-term investments. Cashflow is a measure of the day-to-day capabilities of how well the most liquid of assets is performing. If a business plan involves investments in property, buildings, and major equipment, the time frame will be well beyond a year. What then needs to be considered are long-term investment issues such as depreciation, interest rates, tax deductions, tax credits, etc. This should be discussed with the guidance of a CPA or tax attorney. Long-term investments involve interest not only for obtaining capital but also for long-term financial assessments such as discounted net present value (NPV) and internal rate of return (IRR).

Goodness of fit of actual and predicted cashflow as measured by R^2 can now be used in this application. Both projected and observed cashflows converted to their natural log follow a linear function. This is outlined in the last two columns in the previous exhibit and is displayed in Fig. 7.4.

The new regression formula is:
LN(Y) = a + b(LN(KX))
Multiple R = .84
R^2 = .71

The R^2 is a good fit in explaining the amount of variance of the observed cashflow data explained by model. Very few business plans perform exactly as originally laid out. If the R^2 is performing well below 40%, then the budgeting and cashflow projections in the business plan should be revisited.

All forecasts discussed so far should be monitored as events begin to unfold. It is an education process that never ends. It is statistical analytics and attention to detail that will enhance the ability of any business to anticipate potential risks and plan for successful outcomes. There's an old saying in business and science: "If you can measure it, you can manage it." Any meanigful measurement is always presented in the form of statistics. Growing a business without the use of statistics can diminish the competitive edge so vital for survival. For small businesses statistical tools and analysis help banks make decisions on whether to offer loans, loans that can be used to grow an existing business or start a new one. Statistical analytics allows businesses to measure performance and identify trends.

Part IV
Market Research Principles

Abstract In this section is an overview of market research planning. Regression is also used to determine the geographic reach of a small business usually referred to as the local market area, not only the reach but sales generation by zip code areas. The next phase is to connect sales with customer satisfaction using Likert scales.

Chapter 8
Market Research

Market Research Planning All businesses need information to guide decision-making. Managers trying to understand changing business environments need useable information at the right time. The commercial world is inundated with all kinds of data. That data needs to be converted into useful information so the business manager can make thoughtful decisions. Fundamental to the process is to use statistical methods to gain insight of current operations and future trends.

Managers of companies build a picture of their markets in their mind. They feel that they know what is going on better than what any outsider can tell them. There is truth in the value of experience. But there can be significant prejudice and resistance when the term research is used; "it is not practical." The key in statistical analytics is to make unbiased measurements in a timely and efficient manner.

Commercial success is dependent more than ever, on technological superiority, but not without a better understanding of customers' needs and desires. The need of researching customer attitudes is obvious. For example, launching a new product, what is needed to know is how customers will react. Will they like it? Will they buy it? How much will they pay? How much will they buy? What will trigger their purchase? Launching a product without this information but basing it on internal hunches and opinion (usually optimistic) could be a disaster. Properly conducted statistical analytics is the safety valve for that unbiased look at data-driven decisions.

Market research is an organized effort to gather information about targeting markets and customers: knowing about them, starting with who they are. It is a very important component of business strategy and a major factor in maintaining competitiveness. Market research helps to identify and analyze the needs of the market, the market size, and the competition. Its techniques encompass both qualitative

Supplementary Information The online version contains supplementary material available at https://doi.org/10.1007/978-3-031-40821-2.

techniques such as focus groups, in-depth interviews, and sociodemographic assessment, as well as quantitative techniques such as customer surveys.

Market research, marketing research, and marketing are somewhat related; sometimes these terms are incorrectly used interchangeably. The field of marketing research is much older than that of market research. Although both involve consumers, marketing research is concerned specifically about marketing processes, such as advertising effectiveness and salesforce effectiveness, while market research is concerned specifically with markets and its consumers. Two explanations given for confusing market research with marketing research are the similarity of the terms and also that marketing research is a subset of market research.

Both market and marketing research can be very expansive disciplines to master, but for the manager of a small business, a practical approach is to concentrate on market research. Market research is a way of getting an overview of consumers' wants, needs, and beliefs. It can also involve discovering how consumers perceive a business and how they relate to it. The research can be then used to determine how a product can be promoted using marketing research. Market research is a way that producers in the marketplace study the consumer and gather information about the consumers' needs. Market research starts with sampling business account data, a reflection of current and past business activity. Also required is the ability to survey customers.

What is practical for the small business manager? A good starting point is to look at what is readily available. Almost all businesses use fairly sophisticated computerized accounting systems. They all have the ability to generate reports that summarize customer activity as it relates to the bottom line. Those reports can also be used to generate data for more sophisticated analysis. In order for account data to be practical in conducting market research, it should be structured such that both economic and customer data can be integrated.

Regression and Local Market Area   Regression is very valuable in adjusting predictions using categorical adjustments for various demographics factors such as geographic region and other characteristics such as gender. Geographic region can also be a surrogate for income distribution which is readily available from government published data. For example, there is available through census data sources that break down income tax collections or median income by zip code. Such data combined with a company's sales data is a good indicator distribution of the socioeconomic characteristics of a business' customer base.

The compilation of this type of socioeconomic dataset requires time, but gaining insight of the customer base is invaluable. The first step is to set up a dataset from a sales account with links to customers' zip code. A summary of sales data by individual zip code would be very unyielding to interpret. The first step is to group zip code into broader zip code areas. This is where background research of census data and a map to develop relevant zip code areas (Table 8.1).

The sales and sociodemographic research resulted in condensing the data into four geographic areas. This number was chosen to simplify the presentation of the concept of using regression to project sales by geographic area. It is most likely that many more areas would be of value, especially for larger businesses. From a statistical data perspective, zip codes are categorical data.

Table 8.1 Sales Data by zip code broken down by four geographic areas

Zip_Code	Zip_Area	Sales
0017	1	3,102
0016	1	6,658
0008	1	6,514
0012	1	10,000
0007	1	6,000
0005	1	9,634
0011	1	12,666
0003	2	35,548
0006	2	40,508
0020	2	60,074
..	..	..

The regression results:

	Coefficients	Standard Error	t Stat	P-value
Intercept	-$30,885	$6,135	-5.0	0000
Zip_Area	$47,049	$2,540	18.5	.0000

$$R = .86$$
$$R^2 = .74$$
$$Sales = -\$30,885 + \$47,049 \times Zip\,(\text{Geographic Area number})$$

Used Excel pivot table to summarize sales by zip code area

Area	Sum of Sales
Zip Area 1	$185,228
Zip Area 2	$446,808
Zip Area 3	$608,160
Zip Area 4	$602,265
Grand Total	$1,842,461

As part of small business forecasting, it is key to get a picture of the possibilities for selling products or services in a local market. Looking at local markets will provide information about the types of individuals who might buy products or services and how extensive is the company's geographic reach and what is the competition within the various market areas.

Create a Customer Profile Next there is a need to determine who are the people who will buy products or services.

What age are they?
What is their income level?
What is their education level?
What kind of jobs do they have?
What do they like to do for entertainment?

It may be too cumbersome and difficult for a small business to survey for such data. But a small demographically diverse focus group is a proven way to measure customer opinions. It is set up in guided or open discussions about new products or current views of the company to determine reactions that can be expected from a larger population. The use of focus groups is a market research method that is intended to collect data through interactive and directed discussions by a researcher. If there are issues with lagging sales that don't respond to standard means of marketing, then arranging for a focus group may be what is needed.

Obtaining Customer Ratings A Likert scale is a scale commonly involved in research that employs questionnaires. It is the most widely used approach to scaling responses in survey research, such that it is often used interchangeably with rating scale, although there are other types of rating scales.

The scale is in a format in which responses are scored along a range. When responding to a Likert item, respondents specify their level of agreement or disagreement on a symmetric agree-disagree scale for a series of statements. Thus, the range captures the intensity of one's feelings for a given item. The Likert scale has found widespread use in business and marketing, primarily because of its simplicity.

A scale can be created as the simple sum or average of questionnaire responses over the set of individual items. Likert scaling assumes distances between each choice on the sale are equal. The design of a set of scale items is such that they are highly correlated but also that together will capture a full range of customer preferences.

A Likert item is a statement that a customer respondent is asked to evaluate by giving it a quantitative value based on a level of agreement/disagreement being the dimension most commonly used.

The format of a typical five-level Likert item, for example, could be:

- Strongly disagree
- Disagree
- Neither agree nor disagree
- Agree
- Strongly agree

It is a bipolar scaling method, measuring either positive or negative response to a statement. Sometimes an even-point scale is used, where the middle option of "neither agree nor disagree" is not available. This is sometimes called a "forced choice" method, since the neutral option is removed. The neutral option can be seen as an easy option to take when a customer is unsure, but there is much discussion if it is a true neutral option or just when the respondent is confused. It has been discussed that there is not a significant difference in the use of forced neutral or not.

Likert scales may be subject to distortion from several causes. Respondents may:

- Avoid using extreme response categories (central tendency bias), especially out of a desire to avoid being perceived as having extremist views.
- Agree with statements as presented (acquiescence bias), who may by an eagerness to please.
- Disagree with sentences as presented out of a defensive desire to avoid making erroneous statements and/or avoid negative consequences that respondents may fear will result from their answers being used against them.
- Try to portray themselves in a light that they believe the examiner or society to consider more favorable than their true beliefs.

The biases listed become paramount when questioning individuals regarding socially or highly personal issues. But for the small business manager, the dimensions are primarily three dimensions of primary concern:

1. Overall customer satisfaction with interaction with the business.
2. Satisfaction of value of goods or services based on price and other perceived values.
3. Likelihood of repeat business and recommend to others.

For the small business, these scales are a straightforward indicator of customer satisfaction with the business. They do not deal with more highly charged social and personal assessments. The issue of bias is not a major concern.

Question Design After the questionnaire is completed, below are things to keep in mind when formulating individual questions:

- Make the questions very specific. Notwithstanding the importance of brevity and simplicity, there are occasions when it is advisable to lengthen the question by adding clarification. For example, it is good practice to be specific with time periods.
- Avoid jargon or shorthand. It cannot be assumed that respondents will understand words commonly used by people in the business. Trade jargon, acronyms, and initials should be avoided unless they are in everyday use.
- Steer clear of sophisticated or uncommon words. A question is not a place to score literary points, so only use words in common parlance. Colloquialisms are acceptable if they will be understood by everybody.
- Avoid ambiguous words. Words such as "usually" or "frequently" have no specific meaning and need qualifying.
- Avoid questions with a negative in them. Questions are more difficult to understand if they are asked in a negative sense. It is better to say "Do you ever ...?", as opposed to "Do you never ...?
- Avoid hypothetical questions. It is difficult to answer questions on imaginary situations. Answers may be given but they cannot necessarily be trusted.

- Do not use words which could be misheard. This is especially important when the question is administered over the telephone. For example, fifteen and fifty can sound very similar.
- Desensitize questions by using response bands. Questions which ask about age is best presented as a range of response bands. This softens the question by indicating that precision isn't necessary and only a broad answer is needed.

Forecasting Economic Potential There is a treasure trove of economic information for market research that is contained in business accounting systems. But to unlock that, information requires careful planning. Auditmetrics experience is that standard accounting system reports are useful, but the level of detail may not be sufficient. In a business assessment concerning the lack of growth of a new product, the immediate response of the business manager is usually "let's increase the advertising budget," a typical marketing research response. A review of product sales data for the period of time before and after the initiation of the product was conducted. A random sample was selected to pull records to interview employees and customers. The results indicated staff were not familiar with the requirements of the new product that unfortunately led to customers to be confused. In an audit it was found that there was an unusually high number of product returns. More training of employees is what helped in increasing sales. Fortunately, this assessment was done quickly enough to avoid the company having an image problem.

Routinely Sample Accounts to Monitor Business Activities Many business managers complain that selecting account samples takes a lot of time. It is better to sell rather than sample. There is nothing wrong with pushing for more sales as long as there are not some unforeseen barriers.

What helps is to make inroads on the time issue. Regular timely random samples allow the business manager to deal with small workable subsets of account data representative of the total book. There is no need to use the gold standard of 3% which would require a fairly substantial random sample. The other time saver is the Auditmetrics AI assistance software feature. This makes it possible to rapidly generate a random sample by simply deciding on the desired margin of error and test different number of strata. The business manager can make more rapid forecasts using MS Excel in effect putting the business on a monitor.

Account Data and Customer Input In this section Likert customer rating is linked to customer sales data. It was designed as a preliminary small sample looking at linking customer ratings on price and satisfaction and whether they would recommend the business to others. The random sample was derived from a QuickBooks report that list total sales by customer (Table 8.2):

Please refer to Appendix 2 on the methodology to convert QuickBooks reports into a matrix data file.

The variables are:

Zip_Area—That is the customers' zip code aggregated by areas based on census data that delineate areas of different socioeconomic characteristics based on median family income.

Table 8.2   Likert average Likert scale rating and total sales by zip area

Zip_Area	Lprice	Lsatisfied	Lrecommend	Total Sales
3	5.45	5.77	2.06	$3,102
3	5.56	5.78	2.10	$3,102
1	1.75	3.50	4.05	$4,653
1	1.00	5.47	1.18	$4,653
3	3.33	4.25	4.37	$5,273
3	3.01	4.80	1.80	$5,273
1	1.70	1.77	4.01	$6,000
1	2.89	3.95	2.31	$6,000
1	1.24	4.50	1.36	$7,824
1	3.84	2.80	3.86	$7,824
"	"	"	"	"

Lprice—Average Likert customer rating evaluating price and value of the product
LSatisfied—Average rating for satisfaction with interaction with business staff
LrRecommend—Average rating for repeat business and recommend to others

Average Likert rating is used even though there are scaling issues discussed previously. This assessment is intended to be a quick look at consumer opinions because time is also of value to the customer. A more detailed analysis can be done with a focus group recruiting customers for a more detailed assessment. It usually involves offering some sort of compensation not necessarily cash but product or service discounts as an alternative.

The next step in the analysis is to conduct regression analysis with zip area as the dependent variable and the Likert ratings as the independent variables with the following results:

$$\textbf{Multiple R} = 0.46$$
$$R2 = .21$$
$$N = 60$$

The correlation coefficient is .46 with 21% of the variance of zip area explained. As expected, data variables that measure people's attitudes are not as predictable as the prior regressions involving account data. So as an ongoing prediction model goodness of fit, this provides a preliminary look. Below is the table that gives each independent variable's alpha error. The decision is if the observed coefficients occur by random chance alone is less than 5%, then one can assume there is a measurable effect of that scale by zip area. The Likert rating for price does differ among the different zip areas. Though this is a preliminary snapshot, the observed alpha error (p-value) is so small that it should require further analysis (Table 8.3).

Zip Area 1 is a geographic area which is in the lowest end of median family income based on census data. A breakdown of average sales by that zip area is also at the low end. It may be wise to have an advertising campaign specifically tailored

Table 8.3 Statistical significance of Likert ratings

	Coefficients	Standard Error	t Stat	P-value
Intercept	1.41	0.32	4.41	0.000
Lsatisfied	0.04	0.06	0.72	0.472
Lrecommend	-0.03	0.06	-0.42	0.674
Lprice*	0.21	0.06	3.57	**0.001**

P < .01

Table 8.4 Average sales by zip area

Zip_Area	Average Sales	n
1	$3,206	18
2	$4,890	20
3	$4,601	22

to those families in Area 1. That is the area that may increase demand if price discounts are offered. There can be repercussions if sociodemographic data is used, so further assessment should be conducted. If it happens that this geographic area has an older population, mostly retirees on fixed incomes, then a senior discount for all customers is appropriate (Table 8.4).

Total Process Overview The overall process in conducting forecasting and market research is to:

1. Start with a random sample of accounts.
2. From there use regression to project revenue and expenses.
3. Also add to the account data pertinent variables such as geographic and sociodemographic data.
4. Set up a mechanism to obtain customer ratings using Likert scales.

Step 4 should be part of a total package to obtain customer loyalty. To truly build this loyalty, companies need to move from transaction interaction with their customers to building company customer relationships. The first step in building these relationships is engaging with customers beyond basic one-way dialog. Customers don't feel valued when it takes undo time to contact the business they patronize. At the same time, sending out mass text messages without a prompt response will also not give customers a satisfying feeling either. Correct proactive outreach can help organizations maximize productivity, customer satisfaction, and contributions to the bottom line.

Though much discussion in this book involved quantitative measurements structured to act as part of a business performance monitoring process. Is it worth it? The quantitative methods will expand control of day-to-day operations. Also, when seeking funding for current operations and new business plans, the quantitative methods discussed follow both AICPA and IRS standards at a level of statistical sophistication that usually is available only to large corporations.

Appendices

Appendix 1: The Statistical Method

Statistics is about solving practical problems by collecting and using information derived from data. The aim of this Appendix is to study in more detail statistical methods for analyzing data primarily as they relate to the statistical audit and forecasting. There are three main goals for statistical analysis:

Goal I—Descriptive Statistics is to display and make sense of data using common measures and displays and how it can be organized in databases.

Goal II—Inferential Statistics is to generalize from a sample to a total population that could not be examined in its entirety. The issue of probability and statistical estimation are discussed.

Goal III—Model Building is to develop a model of real-world processes. Statistical theory is used to develop a probabilistic model that best describes the relationship between dependent and independent variables. This is a discussion that will provide the business analyst with an expanded set of very useful and powerful statistical methods for projecting and managing revenue and expenses.

Goal I Descriptive Statistics

Scales of Measurement Descriptive statistics are used to describe, present, summarize, and organize data, either through numerical calculations, graphs, or tables. Descriptive statistics allows a business analyst to quantify and describe the basic characteristics of a dataset. As such, descriptive statistics serve as a starting point for data analysis. One of the biggest advantages of descriptive assessments is that it allows one to analyze facts that helps in developing a more in-depth understanding of a business problem.

A good descriptive statistic is a summary that provides the important quantitative and qualitative features of a collection of data, usually from a sample but not always. Descriptive statistics is distinguished from inferential statistics by its aim to summarize and understand data rather than to make decisions. Generally descriptive statistics, unlike inferential statistics, is not selected on the basis of probability theory. When sample data is used to make inferences, sample descriptive statistics are employed to provide necessary background information.

Scales of measurement is how numerical data elements are defined and categorized. The major breakdown is into qualitative and quantitative scales. Under this breakdown there are four common scales of measurement: nominal, ordinal, interval, and ratio. Nominal and ordinal are subsets of qualitative data used to develop graphs, charts, and tables. It is categorized based on properties, attributes, labels, and other type of identifiers. Interval ,and ratio scales are subsets of quantitative data which include quantities such as dollars, time and weight.

Qualitative data is broken down into categories:

Nominal Scale

- A nominal level of measurement deals strictly with qualitative Data. Observations are simply assigned to predetermined categories.
- Examples: account type; gender; and class membership, i.e., race, sociodemographic, and color of eyes.
- Key Characteristics: Can't perform mathematical operations on the categories. No meaningful ranking of the categories.
- However, one can count the number of observations within each category which can be manipulated mathematically.

Ordinal Scale

- Ordinal Level of Measurement: Ordering of data with the added feature that it can be listed from highest to lowest.
- Examples: Rating individual students based on class rank, rank accounts, or invoices based on dollar volume. Also ranking of categories, e.g., high-, middle-, and low-income individuals.
- Key Characteristics: Can't perform mathematical operations on the categories, but there are mathematical procedures when individuals are ranked in order or the counts within each category.
- It can be considered at times quantitative or qualitative, whether the ranking is individual or category.

Interval Scale

- The interval level of measurement allows for quantitative comparisons among data, but with some restrictions.
- The lack of a true zero value prevents the use of multiplication and division in comparisons.
- Classic Examples: Temperature in degrees C and F; IQ scores
- Key Characteristics: Can use for quantitative assessments. Differences between values can be measured, but multiplication and division do not yield useful information because a true zero value does not exist. For example, temperature scales are arbitrary ranges between the boiling and freezing point of water where zero has no quantitative meaning; it is just another point on the scale. The same is true for the IQ test. The IQ score is a standardized quantitative measure with an average set at 100. One cannot conclude an individual's score of 100 represents twice the intelligence of a score of 50.

Ratio Scale

Effect of True Zero Point

- For example, one can't say that 20 °C is half as warm as 40 °C or that 20 F is half as warm as 40 F,
- However, the Kelvin scale does have a true zero value. At 0 °K (absolute zero −273.15 °C), there is an absence of heat (kinetic energy). Kinetic energy of 400 °K is twice that of 200°.
- Moving from one scale to another:
    - Ratio /Interval -> Ordinal -> Nominal
    - Example: Comparing different companies' net income company actual dollars -> net income ranking of companies, e.g., first, second, third, etc. -> Those showing profit vs those a loss

It is always possible to go from ratio/interval scale to ordinal then to nominal scale:

- Qualitative
    - Nominal Scale
    - Ordinal Scale
- Quantitative
    - Interval Scale
    - Ratio Scale

Descriptive Statistics Uses Summaries collected to explain patterns and monitoring measurements. Many accounts generate large volume of data, and to make sense of them, summary statistics and other methods explain patterns in the data. This provides a method of understanding overall tendencies contained in the data. It is also an important first step towards the implementation of inferential statistics.

Levels of descriptions:

- Describe the size and scope of the data.
- Describe the center of the data.
- Describe the spread of the data.
- Assess the shape and spread of the data distribution.
- Compare data from different groups.

Descriptive Statistics Summarization Types

Account analysis can generate a large amount of data, and to make sense of that data, analysts use statistical summaries and other methods that summarize the data, providing a better understanding of overall tendencies. The major types of descriptive methods that can be performed in Excel:

- Sorting/grouping tables/graphs/ illustrations/visual displays
- Summary statistics, i.e., mean, standard deviation, etc.
- Excel's "=Frequency" function
- Excel's pivot table

Summary statistics make sense of data and are critical in the development of inferential statistical methods. There are two summary statistics procedures important in statistical inference:

- Measures of central tendency
- Measures of dispersion

Measures of central tendency can be divided into three types: mean, median, and mode. The mean is a mathematical weighted average of all transactions in an account book. Mean is the average or total dollar value divided by the number of data points. It has a number of useful statistical properties such as its use in inferential statistics. However, it can be sensitive to extreme scores, sometimes referred to as outliers.

Arithmetic Mean Defined as being equal to the sum of every observation divided by the total number of observations. Symbolically, if a dataset consisting of the values X_1, X_2, ..., X_n then the arithmetic mean, represented by the Greek letter mu and defined by the formula:

$$\mu = \sum X_{i/N}$$

Where:

μ is the mean account dollar value.

N is the total number transactions in the audit population.

X_i is dollar value for ith observation where i can vary from 1 to N.

The mean as a measure of average does have limitations. Other measures of central tendency in certain circumstances may be preferable to the mean. For example, most income summary statistics use the median as a fair representation of the distribution of income and other financial data. The mean is affected by extreme values. This in turn can give a false impression of financial average. For example, if you obtain a sample of middleclass individuals, it would be expected to reflect national statistics. But if for some reason a billionaire is included in the sample, the mean would be weighted by that extreme value.

The median is another measure of central tendency. The median is the value separating the higher half count of a data sample, population, or probability distribution, from the lower half. In simple terms, it may be thought of as the "middle" value of a dataset not based on quantity but the count of each data point. For example, in the dataset {1, 3, 3, 6, 7, 8, 9}, the median is 6, the number in the sample. It cuts the number of observations into the upper half and lower half. This is also referred to as the 50th percentile.

Table A1.1 is a review of the total dollar value of client contracts signed by a consulting firm. The mean is the total dollar value of the contracts $3,410,000 divided by 25 equaling $136,400.

If management wants to impress investors, then it could claim the average contract dollar value is $136,400. It is a reflection of an average contract value, but it is misleading. The 2.25 million contract is 14 times higher in value than the next contract in order. The mean is weighted by one very large contract. With the median, which is 40k, half of the contracts are above and half below this value. This provides

Table A1.1 Contract central tendency values

Contract Value (x1000)	Number Contracts	Central Tendency Measures
$2,250	1	
$156	1	Mean 136k is weighted average
$102	2	
$64	1	
$52	3	
$45	4	
$40	1	Median 40k, 12 above 12 below
$30	12	Mode 30k, most frequent
Total	25	

a clearer picture of the typical contact, especially since the 2.5 million dollar contract is a rare event.

Mode is another measure of central tendency which is the most frequent number in a distribution. It is the value that is most likely to be sampled.

Measures of central tendency provides a concise measure that can be used as a single value to describe a set of data. But a center point alone is incomplete. What was pointed out in the firm's contract size exhibit, the spread of the data, is also an important consideration. How useful would contract mean or median be in summarizing future contract sizes? Not very useful because a spread of the data is an important consideration.

Measures of central tendency:

- The "average" score—total score divided by the number of scores.
- Has a number of useful statistical properties, especially for inferential methods.
- Sensitive to outlier extreme cases.
- Other measures can give an idea of what is the "typical" value in a distribution.
- Mode: the most frequent value in a distribution.
- Median: the midpoint or mid-score in a distribution.

    - (50% observations above/50% observations below)
    - Insensitive to extreme scores

The mean's value in statistical inference also requires a measure of data spread. A measure of central tendency can be used as a prediction of future outcomes, but it is the spread of the data that impacts the prediction's precision.

Standard deviation:

- Is data spread measured by how far or close individual observations are from the mean
- Valuable measure for statistical inference
- Several groups with identical means can be more or less diverse

Range as a measure of dispersion:

- Distance between the highest and lowest scores in a distribution
- ssSensitive to extreme scores
- Compensate by calculating interquartile range (distance between the 25th and 75th percentile points) which represents the range of scores for the middle 50% of a distribution

Calculating Standard Deviation

A simple set of data—0, 25, 50, 75, 100—is used to examine the interplay of standard deviation and variance:

Start with the mean:

Table A1.2 The value of
variable X, X deviation form
the mean and
deviations squared

X	X-μ	$(X-\mu)^2$
$0	-$50	$2,500
$25	-$25	$625
$50	$0	$0
$75	$25	$625
$100	$50	$2,500
Σ=	$0	$6,250

$$\mu = \Sigma X_i / N = \$50$$

Then compute deviations ($X_i - \mu$) from the mean and square deviations (Table A1.2):

The mean as a weighted average results in deviations above the mean canceling out those below the mean, so the sum is always zero. Therefore, the standard measure of spread around the mean is a two-stage process. Each deviation is squared which eliminates the net zero result of summing deviations.

Step 1 Measure Variance

Variance is an average of squared deviations:

$$\sigma^2 = \Sigma (X_i - \mu)^2 / N$$
$$= \$6250 / 5 = \$1250$$

Variance is in itself is useful in statistical inference. For example, there is a statistical analytical method called analysis of variance (ANOVA). It's used to test differences among means of several random samples. It is also a measure of the efficiency of the regression mathematical model. These are analytical methods that will be discussed later.

Step 2 Measure Standard Deviation

To be understandable, squared units are reduced to the original scale by taking the square root of variance. This provides a measure of the spread of data in relation to the mean. It can also be used in calculating the coefficient of variation ($COV = \sigma/\mu$) which is a standardized measure of dispersion of a frequency distribution. A COV can be calculated for any given quantifiable data. Unrelated dataset COVs can be compared to one another in ways that other measures cannot.

Standard Deviation

$$\sqrt{\sigma^2 = \sum_{N}^{i=1}\left(X_i - \mu\right)^2 / N}$$
$$= \sqrt{\$1250} = \$35.4$$

Goal II Inferential Statistics

The Nature of Random Samples
Statistical account sampling is the basis to understand a total account when it cannot be measured directly only inferred.

Figure A1.1 exhibit outlines the random sample process.

Frequently there is a need to know if two or more accounts differ from one another with respect to a parameter they share. These comparisons generally involve using sample statistics to estimate the value of account parameters.

- **A parameter is a number that describes the population**: It is a fixed number in practice and is generally unknown. With computerized business systems parameters of interest, such as the mean or account total, can be determined by system reports. Auditmetrics software uses this fact to determine the validity of sample statistics.
- **A statistic is derived from a sample**: It is a quantity that is calculated from a sample. It is a random variable that has a specific distribution function, for example, the normal curve. Sample statistics are used to make inferences about population parameters.

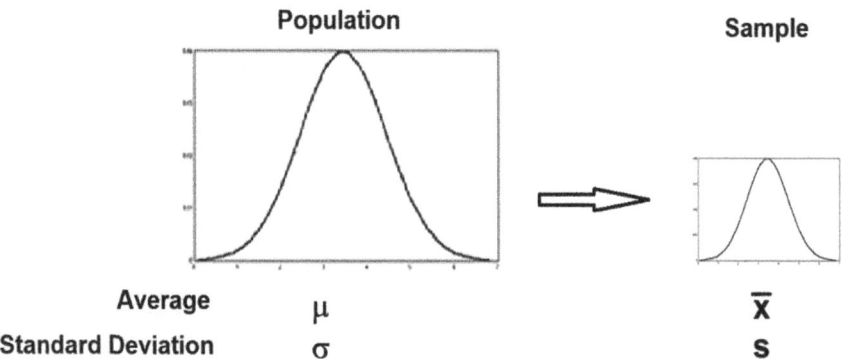

Fig. A1.1 The process of selecting an account sample as an unbiased subset of the population

- **Random sample variables:** $X1$, $X2$,..., Xn account data are said to form a random sample of size n if the Xi's are independently selected and each Xi has the same probability of being selected.

Sampling in Business Sampling is a technique of selecting a subset an account, so one can use statistical inference methods to estimate characteristics of the total book. If a business would like to research the customer demand for a new product, it is impossible to conduct study involving everyone. The market researcher will use a sample of a firm's customers to provide information for rating the new product. This is the same for an auditor to determine how many account transactions are in error due to system breakdown or human error.

Probability Sampling Probability sampling refers to the selection of a sample from a population, when the selection is based on the principle of randomization. Probability sampling is more complex, more time-consuming, and usually more costly than non-probability sampling. Probability sampling allows analysts to create a sample that accurately represents a population of interest.

Convenience Non-probability Sampling Some business researchers choose audit data on the basis of what elements are easy to obtain. This sampling method is not a fixed or predefined selection process. This fact makes it difficult for all elements of a population to have equal opportunities to be included. A business manager for a week asks customers to rate their service as they leave. This may provide valuable insights as to potential opportunities or problems. As indicated by its name, convenience sample data is easily accessible. But there is a price to pay for this ease of effort: convenience samples are virtually worthless in statistical theory.

The reason a convenience sample cannot be used for statistical analysis is lack of the assurance of representativeness of the population. Selection all of the customers for one week may share some common characteristics; it cannot be assumed they represent all customers of record. The traditional block sample method of sampling for the statistical audit is an example of this type of sampling. In this method an auditor may ask for all invoices for the month of February, June, and October.

There are two types of probability sampling techniques relevant in sampling business accounts:

1. **Simple Random Sampling:** This is an easy probability sampling technique that saves time and resources. It is a reliable method of obtaining information where every single member of a population has an equal chance of being selected, but it comes with statistical inference efficiency problems especially when the population is skewed.
2. **Stratified Random Sampling:** This is a method in which the auditor divides account dollars into separate aggregated groups that don't overlap. They are organized to draw a random sample from each group separately. It is the inherent sampling technique used by Auditmetrics.

Benefits of probability sampling:

- **Reduce Sample Bias:** Using the probability sampling method, the bias of the sample is minimized.
- **Reproducibility:** Probability sampling can be consistently replicated by a team of auditors with results that can be directly compared.

Sampling and Data Types There are two basic types of data sampling techniques: variable sampling and attribute sampling. Variable sampling is based on quantitative scales such as weight, length, and dollars. Variable sampling is the standard for sales and use tax audits or in any auditing situation when specific dollar quantities are assessed for errors or compliance with regulations. The Internal Revenue Service (IRS) has established statistical guidelines that are becoming a national standard. The fundamental goal of variable sampling is to answer the question of quantity or "how much?" Appendix IV outlines the IRS's own unique use of confidence interval and probability for allowable deductions and tax credits.

Attribute sampling data are classified categorically. For example, data tracking whether accounts receivable items are past due, they could be classified as a "yes" or "no." Attribute sampling is integral to opinion polling and market research where the pollster seeks the characteristics of targeted subsets of the population. In that context, data stratifications are situation dependent. For example, a pollster or market researcher may be interested in demographic breakdowns of potential customers. There are multiple variations of attributes, but in terms of economic relevance, they should ultimately relate to the variable dollars.

Attribute sampling is classification dependent. For example, an accountant may want to examine customer base in terms of age category or other sociodemographic classification. These are attributes that are mutually exclusive, and the purpose of the audit would be to answer the question "how many?" customers are in each category.

Examples of typical attribute sampling breakdowns are:

- 20 of 100 accounts receivable invoices were past due.
- 10 of 40 inventory invoices greater than $1000 contained a signature.
- 19 of 20 fixed assets purchases had a supporting authorization document.
- 2 of 11 supplier invoices indicated the early payment discount was not taken.
- 13 of 211 journal entries were posted to the wrong account.

The results of an attribute sampling test can be compared to a criterion previously established. If the test results fail the standard, then the accounts should be carefully examined for possible remedies. If a sample estimates 10% of account payable invoices are experiencing delays and errors. It then will be necessary to impose additional controls. The yes/no attribute triggers the introduction of changes such as retrain staff and/or alter invoice management procedures. Also, at the same time, the variable dollar impact should be assessed.

There are times when an accountant can exploit the benefits of both methods of sampling. The auditor may be interested in revenue and cost estimates (variables) by different departments (attributes) of the business or class of transaction.

The quantitative nature of dollars provides more information about each unit of observation and therefore, in statistical terms, a more powerful estimator. An attribute sample is a collection of individual observations based on a common classification. Therefore, each attribute contains a collection of a limited number of counts, therefore lacking a certain amount of specificity rendering projections with less statistical power.

Statistical Audit Variable/Attribute Approach The Auditmetrics system provides a step-by-step methodology that combines the characteristics of both variable and attribute sampling. This combined approach is used by sales tax auditors. The auditor is interested in "how many" transactions in an account are in error, such as sales revenue not properly taxed. But interested is not only in the attribute transactions in error (yes/no) but what is the dollar volume of sales tax lost?

A well-chosen representative variable random sample involving dollars can be broken into any attribute of interest. For example, an auditor wants to determine the number of transactions in an account where there is a failure to pay the appropriate tax:

1. The first step is to set up a variable sampling procedure to obtain a representative random sample based on dollars. Once this sample is drawn, standard statistical tests can be used to determine the validity and reliability of the sample.
2. Because business accounts are commonly processed electronically with software such as QuickBooks®, key economic characteristics of the total book are known. This software provides the means to test whether a sample is a valid statistical subset and not an outlier. If a sample estimate, such as total book value, varies greatly from computerized reports, then the sample validity can be called into question. A new sample should be drawn. There is a myriad of variables that can be tested in this manner including revenue, expenses, tax credits, etc.
3. Once a variable sample passes the validation tests, then the auditor can proceed to use the sample to decide which transactions are in error. The first step is setting up an attribute of those transaction in error or not, yes or no. This dichotomous attribute follows the binomial distribution and can be used in statistical estimation of both the percentage of transactions in error but also their total dollar volume.

Below are other examples of issues relevant to statistical sampling of accounts:

• What is the error rate of paid medical insurance claims which were valid but not paid?
• Do sales summary reports match up with credit card company reports?
• Do bank deposit logs match the bank's reporting?
• How often are invoices voided without explanation?

- Do managers record all supplies pulled from inventory in the inventory log?
- What percentage and dollar value of expenses have been properly recorded for a federal research tax credit?

The examples first ask the question of how many transactions are in error, and if found lacking, then dollar volume impact needs to be assessed. Account transactions do not operate in a vacuum. There are always subsidiary personnel, customer, and organizational issues that should be explored. Take the example of a medical claims adjudication case. In addition to determining the dollar error rate, there should also be an examination of attributes as they relate to specific staff and claim types, medical office structure, and health plan design characteristics. The auditor can assess dollar revenue volume slippage in the system, but a full understanding of relevant attributes should be explored. Attributes provide the basis for the detective work to identify problems.

Normal Curve Thus far the discussion has been about the types of account data used in business analytics. These are the building blocks to make decisions about an account's value. The random sample estimation process, as displayed in Fig. A1.1, is needed to convert data into useful information. Sample statistics such as mean $\overline{X}$ and standard deviation S are used to estimate the population mean μ and standard deviation σ that we cannot measure directly.

To determine the probability how well a sample estimate matches a population parameter, an understanding of the mathematics of the random process is required. The starting point is the normal or the bell-shaped curve.

The normal curve has been attributed to the mathematician Friedrich Gauss. In 1801 astronomers had discovered a dwarf planet Ceres in the Asteroid Belt. Many astronomers tried to predict its orbit but failed. There was a limited amount of observation data of Ceres' orbit. No one could pinpoint the orbit. It was Gauss who exactly predicted its orbit.

His approach involved considering the errors in the measurements of Ceres' position. From error assumptions, Gauss came up with the probability density function (PDF) of the normal distribution. But Gauss did not stop there. He then used a method that he had devised earlier to produce an improved prediction of the position of Ceres. This method is what is now called the method of least squares. This is the basis of regression analysis which will be further discussed in the section on model building.

There seems to be some ambiguity as to whether Gauss actually applied the theory of least squares in computing the orbit of Ceres. Many astronomers struggled to find the new dwarf planet, using the few observations available at the time. All were unsuccessful; the first to succeed was Gauss whose predictions were spectacularly accurate.

The shape depicted in Fig. A1.2 represents the random process. Suppose an account has a mean μ of $100 and a standard deviation σ of $15. The goal is to estimate the mean of the population mean using the mean $\overline{X}$ of the sample. Multiple random samples can be drawn. Then the normal curve can be used to answer the

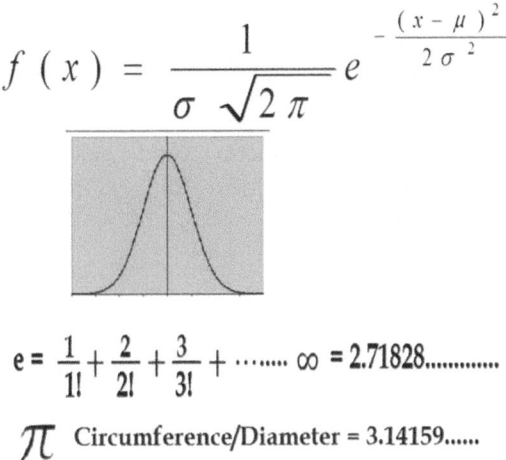

$$f(x) = \frac{1}{\sigma\sqrt{2\pi}} e^{-\frac{(x-\mu)^2}{2\sigma^2}}$$

$$e = \frac{1}{1!} + \frac{2}{2!} + \frac{3}{3!} + \cdots\cdots \infty = 2.71828\ldots\ldots\ldots$$

π Circumference/Diameter = 3.14159......

Fig. A1.2 The mathematics of the Gaussian curve

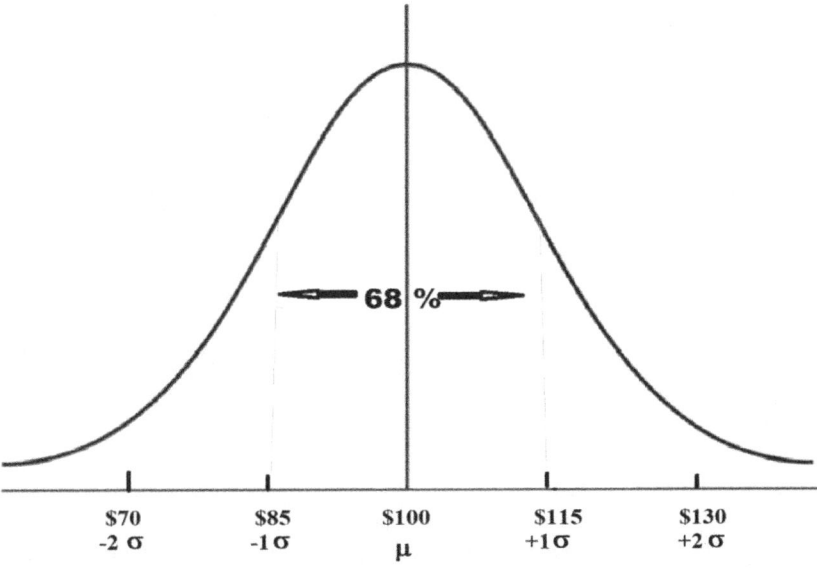

Fig. A1.3 Areas under the normal curve as indicator of probability

question what is the probability a sample mean within one standard deviation of the population mean or $100 \pm $15 or between $85 and $115.

As Fig. A1.3 points out, the probability of a random sample $\bar{X} \pm 1\sigma$ is 68% of the accounts true mean value μ is 68%. An important standard deviation range used in statistical inference is the range $\pm 1.96\sigma$ which represents a probability of 95% as it is used for the 95% confidence interval. The normal curve is best described as a symmetrical distribution where the mean, median, and mode are the same value.

Multiple Normal Curves

A multitude of normal distributions have their own value for mean and standard deviation. The formula displayed in Fig. A1.2 is not useful in calculating probabilities. There is a need to convert an infinite number of normal curves into a standardized normal curve where areas under the curve can be calculated. Below are three very different normal distributions:

X1	$\mu = 100$	$\sigma = 15$
X2	$\mu = 10$	$\sigma = 2$
X3	$\mu = 1000$	$\sigma = 250$

The following formula can create what is known as the standard normal distribution:

$$Z = (X_i - \mu) / \sigma$$

From X_i, the variable of interest is subtracted the mean, and that difference is divided by the standard deviation. Properties of the standard normal distribution is that it will always have a mean of zero and standard deviation of 1. It also has the advantage that areas under the curve (probability) can be calculated using integral calculus techniques.

Drilling down the three distributions to the standard normal (z).

X1	70	85	100	115	130
X2	6	8	10	12	14
X3	500	750	1000	1250	1500
Z	-2	-1	0	$+1$	$+2$
	-2σ	-1σ	μ	$+1\sigma$	$+2\sigma$

X1 to X3 three very different normal curves are converted into the standard normal curve Z. The Z score is an indicator of the number of standard deviations above and below the mean (Fig. A1.4).

Normal curve X1 happens to be the distribution of the IQ test. Is an IQ score of 130 a good score? First let's convert it to its Z score:

$$Z = (130 - 100) / 15 = 2$$

The Z of 2 or two standard deviations above the mean has the approximate percentile rank of 97.5%. This indicates that for an individual with this score, there are only 2.5% who have a higher score. Conversely 97.5% scored below.

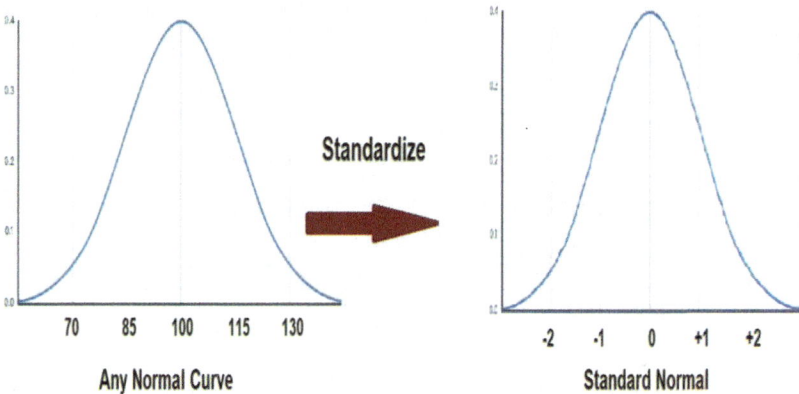

Fig. A1.4 Converting any normal curve into a standardized normal curve

Sampling Distribution A sampling distribution is a probability distribution of a statistic that is obtained through repeated sampling of population with a specific sample size n. It describes a range of possible outcomes for a sample derived statistic. There are two sample statistics that are of importance when sampling accounts, the mean and total dollars.

Population Parameter mean μ Sample Statistic Mean $\bar{X}$
Population Parameter Total τ Sample Statistic Total T

$$\text{Sample mean } \bar{X} = \sum x_i \, / \, n$$

$$\text{Sample Total } T = \sum x_i$$

Case Study 2 covers a method to estimate τ using a sample. Sampling distribution of the mean is used in validating strata validity.

Both the sample total and mean are statistics that have their own distribution properties.

Table A1.3 outlines the relationship between the audit population and sample distribution of the mean. Repeated sample means will have a distribution with a mean that is the same as the mean of the population. It will have a standard deviation that is population standard deviation divided by the square root of sample size n which is commonly known as standard error.

Central Limit Theorem The central limit theorem (CLT) states that if you have a population with mean μ and standard deviation σ and take a sufficiently large number random samples, then the distribution of the sample means will approximate the normal curve. CLT provides extraordinary power in generalizing from a random sample the properties of the population.

Table A1.3 The relationship between the sampled population, sample statistic, and sample distribution

Population to be Sampled	Sample Statistic	Sample Distribution of the Mean $\overline{X}$
Population Mean μ	$\overline{X}$	Mean = μ
Population Std Dev. σ		Std Dev = σ/√n*

* Standard Error

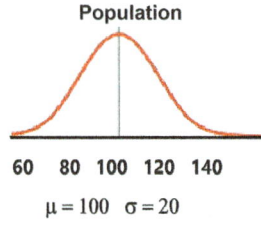

Population

60   80   100   120   140

μ = 100   σ = 20

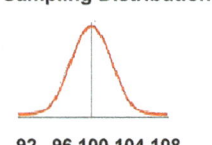

Sampling Distribution

92   96 100 104 108

n= 25 $\overline{x}$ = 100   s = σ/√n = 4

Fig. A1.5 Central limit theorem comparison of population and sampling distribution

1. The sampling distribution of sample means is a normal distribution if sample size is large enough.
2. The mean of the sampling distribution of sample means equals the mean of the population.
3. Th standard deviation of the sampling distribution of sample means equals the standard deviation of the population distribution divided by the square root on n.

Figure A1.5 displays the impact of CLT. Financial data tends to be skewed to the right. Item one from the list of the properties of the CLT states that the sampling distribution is a normal distribution "if the sample size is large enough." When the sample is allocated over several strata, Auditmetrics determines if the sample size is large enough to assure normality.

Hypothesis Testing Hypothesis testing is a procedure where an analyst tests an assumption regarding a population parameter. The methodology used by the analyst depends on the nature of the data used and the reason for the analysis. Hypothesis testing uses sample data to assess the possibility of a hypothesis being valid.

Hypothesis starts with a proposed association and a null hypothesis:

- Null hypothesis (Ho): claim of "no difference between sample and population"
- Alternate hypothesis (Ha): claim of "a difference between population and sample"

In hypothesis testing, an analyst selects a random sample with the goal to test for the two opposite hypotheses: the null hypothesis and the alternative hypothesis. The null hypothesis is assumed, but does the sample indicate the null hypothesis should be rejected and accept the alternative hypothesis that the sample is from a different population?

Table A1.4 from Auditmetrics indicates that all strata accept the null hypothesis except strata 4. For strata 4 the null hypothesis is rejected, and the alternative hypothesis is accepted. Auditmetrics uses the confidence interval test. One can also use the Z test.

Z Test for strata 4:

$$Z = \left(\overline{X}_I - \mu_i \right) / \sigma / \sqrt{n}$$
$$= (\$580 - \$548) / \$115 / \sqrt{61}$$
$$= 2.2$$

Table A1.4 Null hypothesis test comparing sample mean with population mean

Acme Inc.

Strata	Population Pop. Mean	Sample Sample Mean	Sample Sample Std. Dev.	Sample Size	Strata Validity Test Lower 5% Alpha Bound	Strata Validity Test Upper 5% Alpha Bound	
1 0-49.99	$19.02	$17.88	$14.95	47	13.61	$22.16	pass
2 50-174.99	$91.67	$96.83	$32.57	49	87.71	$105.95	pass
3 175-399.99	$258.69	$254.11	$118.76	53	222.14	$286.08	pass
4 400-824.99	$548.66	$580.06	$115.76	61	551.01	$609.11	fail
5 825-1600	$1,106.04	$1,121.28	$423.85	78	1,027.22	$1,215.34	pass
6 > 1600	$1,880.24	$1,880.24	$632.25	463			

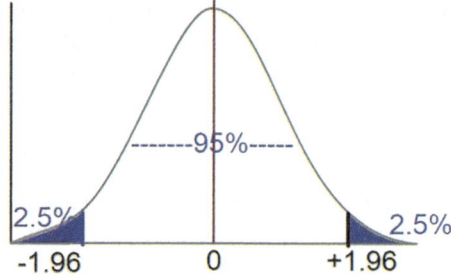

Fig. A1.6 Boundaries for a two-tailed 5% statistical test of significance test

It was previously discussed that a Z score of 2 is approximately the 97.5 percentile. Z equal −2 is approximately the 2.5 percentile. A more accurate percentile value for Z is 1.96. This is boundary as the basis for what is called level of significance.

The significance level (denoted by α) decision rule is generally set at 0.05. There is a 5% chance that the alternative hypothesis will be accepted when the null hypothesis is actually true. The 5% is also called alpha error. The computed Z is 2.2 which is greater than the 1.96 cutoff. The null hypothesis is rejected, and the alternative hypothesis is accepted. This rejection can be erroneous because 5% of random sample means can actually come from the audit population. By convention this is considered a reasonable risk. This error probability is set a priori before implementing the sampling process (Fig. A1.6).

Confidence Interval Test for Strata 4

Confidence intervals provide a range of possible values and an estimate of the precision for a parameter's value. The confidence interval indicates where the population parameter is likely to reside. For example, a 95% confidence interval around a mean suggests one can be 95% confident that the population mean is within the boundaries of the interval.

$$\mathbf{CI} = \bar{X} \pm \mathbf{Z}\left(s / \sqrt{n}\right)$$

$$= \$580.06 \pm 1.96\left(61 / \sqrt{115.76}\right)$$

$$\mathbf{CI} = \$551.01 \leq \$580.06 \leq \$609.11$$

Since the population mean is **$548.66** and is outside the 95% confidence interval, the null hypothesis is rejected.

IRS statistical sampling for estimating deductions and tax credits such as the research tax credit uses a one sided 95% confidence interval. Its Z score is 1.645, not 1.96. IRS has also developed "relative precision" to introduce a sliding scale for confidence intervals. This is discussed in Appendix IV.

Statistical Error Exactly what are the dynamics of this statistical decision process? A simple probability experiment with a coin toss can help clarify. Suppose it is proposed that with the flip of a coin a wager is set up, if a toss is heads, a certain dollar amount will be paid, but if tails the opponent pays. Of course, it is a bet that in the long run should be a net zero for each participant because the probability of heads would be 50%. It is better odds than at a casino which always favors the house. It is the excitement of a long run of good luck or a large wager with a big return.

Suppose a participant raises a question if the game of chance is valid or the coin is a fake specifically weighted to favor the opponent? What is decided is to do a statistical experiment by collecting sample data with the toss of the coin five times. Since the flip of a coin has a dichotomous outcome, 1- heads 0- tails, the binomial distribution can be used to calculate the probability of all of the possible outcomes of five tosses of the one coin (Table A1.5):

An outcome of all heads or tails is 3.10%. If our target alpha error standard is $p \leq 5\%$ and if the outcome after five tosses is all heads, that is rare enough for a conclusion that the coin may be biased favoring heads. The decision is to reject the null hypothesis of a fair coin and accept the alternative hypothesis, the coin is biased. Can that be a mistaken decision? Yes. It is reasonable to consider a fair coin can turn up heads five times. If the coin turns out to be unbiased, then the conclusion is incorrect? If that happens then one is committing an alpha error. In more traditional statistical terminology, a true "null hypothesis" has been rejected. There is also the possibility of accepting a false null hypothesis.

Misstatements Making a statistical decision always involves uncertainties, so the risks of making errors are unavoidable in hypothesis testing. The probability of making an alpha (α) Type I error is the significance level, while the probability of making a Type II error or beta (β) is influenced by sample size.

Since the Type I error is set in advance of selecting the sample, Auditmetrics sets Type I error at 5%. One can minimize Type 1 errors is by raise the level of statistical significance. A higher level of statistical significance requires a larger sample size. That increases the cost of the audit. When it comes to adjusting sample size, Auditmetrics deals with sample size to increase statistical power and the reduction of Type II error which also has an impact on reducing Type I error.

Table A1.5 Probabilities of five tosses of a coin based on the binomial probability distribution

No. Heads	Probability
0	3.10%
1	15.60%
2	31.30%
3	31.30%
4	15.60%
5	3.10%

Table A1.6 Summary of audit misstatement and correct decisions

Null Hypothesis is	TRUE	FALSE
Rejected	Type I Error False Positive Probability $= \alpha$	Correct Decision True Positive Probability $= 1-\beta$
Not Rejected	Correct Decision True Negative Pobability $-1-\alpha$	Type II error False negative probability $= \beta$

Making a Type II error means there is a failure to detect a difference when a difference actually exists. When that happens, the analysis doesn't have enough statistical power. Statistical power $(1 - \beta)$ is a measure of the ability of a statistical test to detect a true difference of a certain size between the null and alternative hypotheses. It is defined as the probability of correctly rejecting the null hypothesis when it is false. The best control for Type 1 error is to increase sample size. Auditmetrics does an adjustment of strata sample size in order to balance statistical power, sample size, and audit cost (Table A1.6).

Goal III Model Building

In statistics, a model is a mathematical equation that describes a functional relationship between two and more variables. A basic model is the regression model. It is a mathematical equation that relates a dependent variable with one or more independent variables. The simplest mathematical equation used in model building is linear regression.

In essence past data is used to predict future outcomes. The assumption is that the pattern of past data will likely continue in the future. The ability to use regression to model economic situations and predict future outcomes makes the regression model an extremely powerful business tool. The power of regression models contributes to their popularity in economics, business, and finance.

However, markets and business climates can exhibit unanticipated volatility. The key for survival is to constantly monitor data trends. It is usual for a business to see a certain amount of regularity in its data. But markets can suddenly change. If a business constantly samples its accounts, it will have an excellent chance of detecting forthcoming trends before being overwhelmed. Broadly speaking, market volatility measures the amount of deviation away from the average. Volatility is also the rate of change of deviations over time.

The ability of the regression model to accurately predict future trends depends on the strength the relationship between the independent and the dependent variable. The model can only be as good as the quality of the data that is used to build it. The most efficient strategy is to sample accounts on a regular basis. Regular management reports may hide those subtle changes that may be indicators of future unanticipated changes. Regular statistically valid sampling allows for timely and efficient detailed account analysis to find trends that may be hidden in standard accounting computerized reports.

The Linear Model

One way to determine the volatility of the relationship between dependent and independent variables is by computing the correlation coefficient. If the correlation shows a strong relationship (closer to 1, irrespective of the sign), then one can be comfortable with the current model. The gradual degradation of correlation is a sign that other inputs are becoming more important and that the current independent/dependent variable model should be re-examined. This may be the first sign major changes are under way.

The simplest linear model is the *bivariate model*:

- Have a set of data with two variables, X and Y.
- The goal is to find a simple, convenient mathematical function that explains the relationship between X and Y.

Suppose dollars spent on advertising is plotted against increase in sales as shown in the following graphical representation (Fig. A1.7).

This example is a plot of the amount of money spent on advertising (X) with an increase in total sales (Y). This mathematical model is a linear function where b is the slope of the line also called the regression coefficient and a is the Y intercept which is the value of Y when X is zero. In this example all data points follow a straight line. The data fit the linear model perfectly; therefore the correlation (r) is

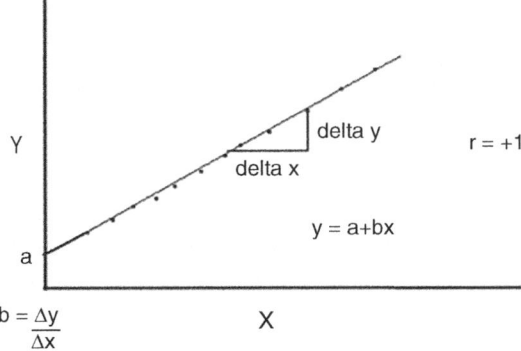

Fig. A1.7 Linear regression with perfect correlation

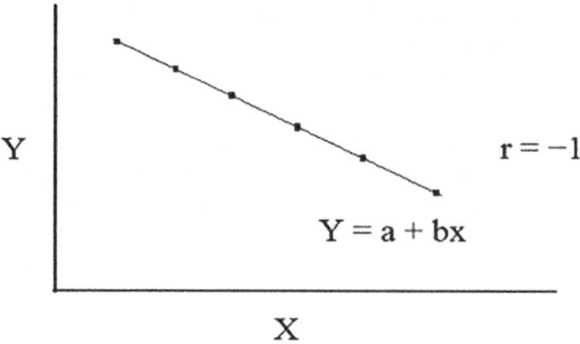

Fig. A1.8 Perfect inverse relationship

1. It is a perfect model with X being the independent variable and Y the dependent variable. If X is known, then X perfectly predicts Y.

This plot is an example of a direct relationship between X and Y, as X goes up so does Y. There can also be an inverse relationship between X and Y. For example, if X is price and Y total sales, the following plot examines this mathematical relationship (Fig. A1.8).

Raise price then total sales go down. This plot is an example of an inverse relationship, but it should be noted the real-world relationship between price and demand can be very complex. It depends on "price elasticity" of demand on how sensitive price is to the demand for a particular good or service affected by the possibility of substitution of other goods and services.

The examples so far reflect a world where the model is perfect. All data points fit the mathematical function perfectly. A complete lack of functional relationship between X and Y look like (Fig. A1.9):

This is an example indicating no functional relationship between X and Y with correlation equal zero. At each level of X, there are both high and low values of Y.

Useful real-world data is somewhere between the previous extreme scenarios. For example (Fig. A1.10):

Most real-world data do not perfectly follow mathematical functions. In this example there is a general trend, as X increases so does Y. A straight line can be fitted to summarize the functional relationship to make projections. However around that line is a zone of uncertainty. The linear function in this example contains an error value. Measuring statistical error was presented previously with the discussion of standard deviation around the mean.

The line depicted in Fig. A1.10 is what is termed the "best fit line." One could manually place a plot of a straight line that appears to be a good fit of the data. Y' is the straight line (Y' = a + bx) which is the mathematical model, while Y is each individual observed data point. To obtain maximum statistical efficiency is to use differential calculus to derive a straight line that minimizes the deviations (Y-Y') or error around the prediction model.

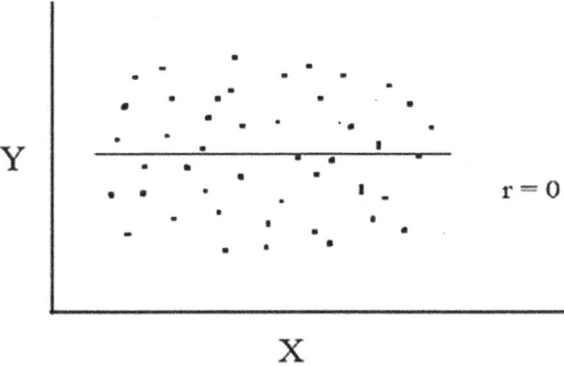

Fig. A1.9 Lack of relationship between X and Y

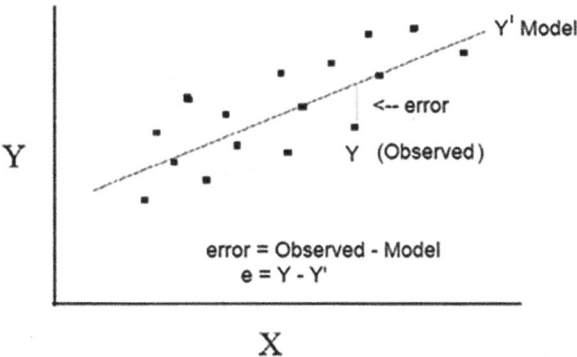

Fig. A1.10 Linear relationship between X X and Y with R = .8

Standard deviation discussed in the descriptive statistics section is a measure of the spread of individual data points around the mean. In this case the measure of spread is data points around the straight line. The error measure (e) outlined in the graph is the deviation of each data point. The key is to find a line that minimizes deviations around the straight line:

$$Minimize(e)^2 = \text{least square line}$$

The formula's goal is to minimize the square of the deviations around the line. Why the square of the deviations? It is the same reason why to first calculate variance around the means since deviations above and below the mean cancel out to zero. With the least square line, all deviations above the line will cancel out those below and add to zero. The formula is:

$$S^2_{x.y} = \frac{\Sigma(e)^2}{n-2}$$

The formula measures the variance around the regression line. Since we will be working with a sample of the data, the formula is the square of the deviations around the line divided by the degrees of freedom (df). In the standard deviation around the mean, degree of freedom is n − 1. When you draw a sample, one observation is needed to fix the mean, and the rest are free to vary. With this formula two data points are needed to fix a straight line. The rest are free to vary.

Mathematical methods of minimizing the square deviations around the straight line are accomplished by using differential calculus minima/maxima calculations. With the regression straight line:

$$Y' = a + bX$$

the formulas for a intercept and b slope to minimize error around the straight line:

$$b = \frac{n\sum xy - (\sum x)(\sum y)}{n(\sum x^2) - (\sum x)^2}$$

$$a = \bar{y} - b\bar{x}$$

We now have the ingredients to start to use regression, sometimes referred to as ordinary least squares (OLS), as a tool in predicting future economic performance based on historical data.

Error Distribution Around the Regression Line The error term for the regression line is based on the assumption that the errors are randomly distributed as displayed below (Fig. A1.11).

- The error term may assume any positive, negative, or zero value and is a random variable.

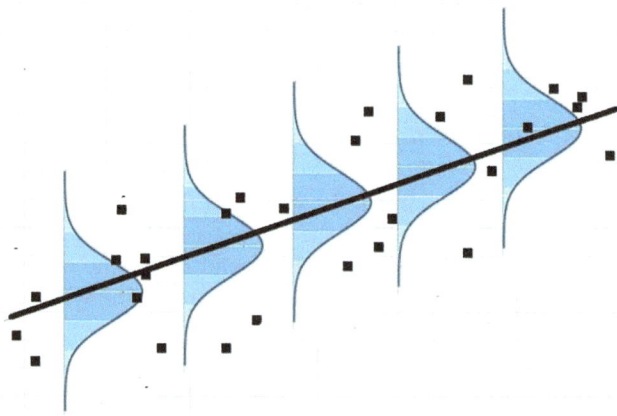

Fig. A1.11 Error term around the regression line following the normal distribution

- The mean value of the error term is zero $(\bar{e} = 0)$.
- The variance $\sigma^2_{y.x}$ of the error is constant at any given value of X.
- The error follows the normal distribution.

Appendix 2: Converting QuickBooks® Standard Reports into Auditmetrics® Data

This appendix deals with converting a QuickBooks Excel report into a format that can be readily read in by Auditmetrics. The basic process is generic and can be used for any Excel file derived from any other computerized accounting system.

QuickBooks Excel Data Conversion

The latest version of Auditmetrics relies on standardized data for input and output. The data standard is text files of two types: .txt and .csv files. Auditmetrics data input is the tab-delimited text file (.txt). The sample data output is both the tab-delimited text file (.txt) and comma-separated value file (.csv) which is also a text file. The (.csv) file has the advantage that it can be readily read in by Excel. It has all the functionality of a spreadsheet and immediately can be saved as a spreadsheet.

Data for Auditmetrics has to be in either a (.csv) or a tab-delimited (.txt) file. These universal file structures make it possible to plug in data from QuickBooks standard reports without the need for specialized software. It works best with reports that detail data at the transaction or invoice level such as sales, expenses, accounts payable, accounts receivable, etc. Reports that detail a summary at a sufficient detail are also useful. For example, accounts receivables per customer (price x quantity) is a useful detail. Below is an example using a "Sales by Customer Report" in the Report section of QuickBooks.

Excel data files converted to text files can become seamless plug-ins to Auditmetrics. However most Excel spreadsheet reports are not specifically formatted as a rectangular data matrix. A data matrix is a rectangular layout where the top row is the variable name. Each column is the individual variable, and each row is the individual record. There are standard QuickBooks reports that can be exported to Excel.

In this exercise *QuickBooks Desktop Pro* is used to export a "sales by customer report." That report is to ultimately create a plug-in for Auditmetrics. QuickBooks can generate reports in many different formats.

Below is a report with conversion options. Select **Excel** as the output for the standard report.

Customize Rept. | Coment on Rept.|Share Template|Print|Email| **Excel** |Hide Header								
Sample Report								
	Type	Date	Num	Name	Item	Qty	**Amount**	Balance
Ecker Design								
	Invoice	12/15/2021	131	Ecker D	Garde	1.00	1.00	1.00
	Invoice	12/15/2021	131	Ecker D	Pest C	1.00	1.00	2.00
Total Ecker Design						2.00	2.00	2.00
Golliday								
75 Sunset Rd.								
	Invoice	12/02/2021	120	Gollida	Plants	10.00	10.00	10.00
	Invoice	12/02/2021	120	Gollida	Install	54.00	54.00	64.00
Total 75 Sunset Rd.						64.00	64.00	64.00
Golliday Sporting G						64.00	64.00	64.00
Heldt, Bob								
	Invoice	12/08/2021	123	Heldt, F	Plants	2.00	2.00	2.00
---------->	Invoice	12/08/2021	123	Heldt, F	Plants	3.00	3.00	5.00
	Invoice	12/08/2021	123	Heldt, F	Fertili	6.00	6.00	11.00

The QuickBooks Report menu bar at the top drop-down list has an option to generate Excel files. One can either generate a new file or add to an existing file. Exporting to a standard report Excel worksheet will have the same layout as a printed report hardcopy. An Excel file for the report can be readily converted to a .csv file which is in reality a text file.

The relevant invoice data is embedded and not readily reachable as data elements for processing by Auditmetrics. There is a report option that will help in selecting invoice data. Once you select the Excel option, you are then asked where to send the report. Select the last option to create a .csv text file. If you are not using QuickBooks but another accounting system, any Excel report can be converted to .csv by using "save as" and selecting .csv.

Send Report to Excel

Would you like to do with this Report?

o   Create a new worksheet

o   Update an existing worksheet

o   Replace an existing worksheet

•   **Create a comma separated values (.csv) file**

The Report option in this exhibit is not to generate a worksheet but to obtain a .csv file which is a text file readable by Excel. It may look like an ordinary spreadsheet, but it is a special text file that can be manipulated and used to create a rectangular text file that can be plugged into Auditmetrics.

Below is the QuickBooks Excel spreadsheet report converted to a .CSV text file:

Type	Date	Num	Name	Amount
Crenshaw, Bob				
Invoice	12/10/2021	FC 8	Crenshaw, Bob	$60.00
Total Crenshaw, Bob				
DJ's Computers				
Invoice	12/15/2021	132	DJ's Computers	$125.00
Total DJ's Computers				
Ecker Design				
Invoice	12/15/2021	131	Ecker Design	$23.00
Invoice	12/15/2021	131	Ecker Design	$68.00
Total Ecker Design				

Open the .CSV file into Excel and now it is easy to manipulate. Sort the .CSV file by Type and Date that are highlighted above.

Type	Date	Num	Name	Amount
Invoice	12/10/2021	FC 8	Crenshaw, Bob	$60.00
Invoice	12/15/2021	132	DJ's Computers	$125.00
Invoice	12/15/2021	131	Ecker Design	$23.00
Invoice	12/15/2021	131	Ecker Design	$68.00
Crenshaw, Bob				
Total Crenshaw, Bob				
DJ's Computers				
Total DJ's Computers				
Ecker Design				
Total Ecker Design				

All of the invoices are bunched together, and now it is a simple matter to remove all extraneous rows and come up with a rectangular dataset ready for analysis.

Add Auditmetrics Required Variables:

Amount, Absamt, Transaction _ ID, and DataSet.

For more detail concerning Auditmetrics required fields, either skip to Appendix VI or go to AuditmetricsAI.com and download "GettinStarted.pdf."

Num	Amount	Absamt	Transaction_ID	DataSet
131	($23.00)	$23.00	3	Run_1
FC 8	$60.00	$60.00	1	Run_1
131	$68.00	$68.00	4	Run_1
132	$125.00	$125.00	2	Run_1

The highlighted columns in gray are required variables for an Auditmetrics text file. It may have originally named SALES, but the name has to be changed to amount. But added are three other required variables Absamt, Transaction_ID, and DataSet. These variable names are required for an Auditmetrics data matrix:

1. **Amount**—The transaction of interest in the analysis.
2. **Absamt**—Absolute value of each transaction. The dataset *must* be sorted in an ascending 3 order. This is to handle credits.
3. **Transaction_ID** —An identifier for each record; it is a record count.
4. **DataSet**—A name to identify this specific dataset.
5. **Primary Key Optional**—If a dataset is from a relational database with a primary key that links the various data tables, it is prudent to include this variable in the audit population dataset to be sampled.

Variables 1 to 4 names must be in the data file and spelled exactly as above; letter case and order do not matter. If not present, Auditmetrics will reject the file. The same is true if Absamt is not sorted in ascending order.

If you need to add or create other variables, it is safest to use letters and numbers with no embedded spaces. For example, if you want to name a variable Run 1, it is best to use Run _1 or Run1. This rule is true for many database and statistical software.

Transaction_ID has value especially when the dataset is the merging of several data sources. For example, if you have two datasets of 1000 each and use MS Access to merge them, then Transaction_ID 1 to 1000 are from the first dataset and 1001 to 2000 are from the second dataset.

The worksheet is now ready to be "saved as" a tab-delimited .TXT Comma variable-separated (.csv) files are standard text files for data transfer, but for Auditmetrics input only tab-delimited file are required for input because with many

.csv files, commas may be embedded in many fields such as the "Name" field listed below:

Name
Williams, Abraham
Rummens, Suzie
Heldt, Bob
Crenshae,Bob
Hughs, Davis

In this QuickBooks report, the name variable is last name comma first name. This comma is not a variable separator and, if left in place, would distort the dataset. Auditmetrics uses only tab-delimited .txt files as data input which can be readily read into MS Excel. Prior versions of Auditmetrics did allow .csv files for input, but it required two passes of the data, one to cleanse stray commas and then use the clean file for data plug-in. It may not be a problem for small datasets, but with an account of five million transactions, it does take a toll on operational efficiency, and no matter how complete the cleansing software, stray commas may sometimes be left behind. Therefore, it was decided to use only tab-delimited files for input.

Appendix 3: Creating a Relational Database

This book relies heavily on accounting data, but it is up to business analysts to include additional data that goes beyond dollars and the bottom line of traditional business accounts. To expand analytic capabilities, other data such as customer ratings and marketing information need to be added into a database. The simplest way to start is to merge account information with secondary qualitative and quantitative data. For example, a simple brief questionnaire asking customers rating their service, suggestions for improvement and new options, etc. This data can be merged with the straight accounting data of dollar and cents.

Larger-scale data sources via census and other publicly available data can also be a valuable source of customer demographic input. Customer demographics is valuable so that business plans can be tailored for specific market segments. A message targeting senior citizens will be different than that for a teenager. The challenge is how to link secondary add-on marketing data with account information.

Having an understanding of database design is very valuable. A relational database is an easily understandable design that is the basis of Microsoft Access. It can be used to creates tables each derived from separate independent files. The relational database management system (RDMS) creates a mechanism that links those individual files into an integrated unified system. It allows one to easily find specific information. It also allows one to sort based on any field and generate reports that contain selected fields from each table. The fields and records are represented as columns (fields) and rows (records) in a table.

With a relational database, one can quickly compare information because of the arrangement of data in columns. The relational database model takes advantage of this uniformity to build completely new tables out of information from existing tables. In other words, it uses the relationship of similar data to increase the speed and versatility of the database.

The "relational" part of the name comes into play because of key field relationships. Each table contains a column or columns that other tables can key onto to gather information. This architecture can create a single smaller targeted table for specific analysis. The following exhibit is an example of how revenue information can be linked to customer data.

The relational database management system (rdbms) creates a mechanism that links individual files into an integrated unified system. With a relational database, you can quickly compare information because of the arrangement of data in tables the relational database model takes advantage of this uniformity to build completely new tables out of information from existing tables. It uses the relationship of similar data to increase the speed and versatility of the database. The "relational" part of the name comes into play because of key field relationships. The relational relationship between customer and orders tables is linked by a key CustomerID (Table A3.1).

Figure A3.1 displays the relational links between three separate access tables. The primary key variable CustomerID links customer table with orders table. CustomerID has a one-to-many relation with orders. This exhibit points out the power of a relational database. The orders table has its own primary key OrderID so that each order has its own unique identifier.

Table A3.1 The relational relationship in two separate access tables with CustomerID as linking key

Customer Table: **Primary Key**

FirstName	LastName	Street	City_Town	Zip	Phone	Cutomer ID
John	Smith	21 Maple	Holly	99999	555-148-3287	100001
Sarah	Brown	3 Oak	Hamlet	99998	555- 396-1899	100002
Judy	Taylor	267 Elm	Mid Town	99997	555-470-8143	100003

Account Table:

Transaction_ID	Sales	Department	Date	etc.	CustomerID
2	216.76	1001	Jan. 10		100001
4	255.02	1001	Jan. 12		100001
5	299.24	1001	Jan. 13		100003
8	405.79	1002	Jan. 15		100002
11	562.08	1002	Jan. 20		100002
15	105.06	1001	Feb. 1		100003
17	201.99	1002	Feb. 6		100003

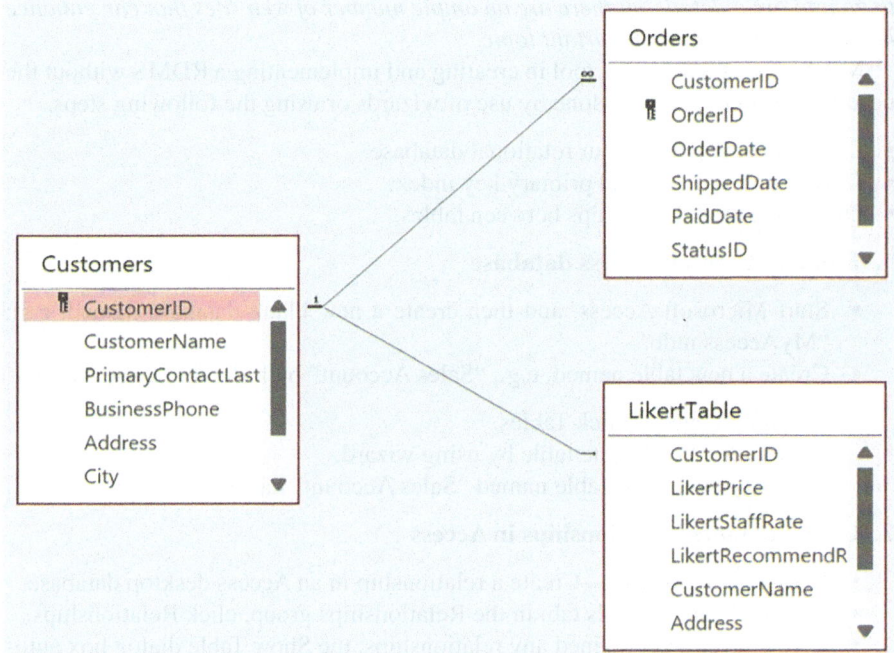

Fig. A3.1 The linkage of externally developed Likert scale data

An Access DB sample of the design is available in the https://auditmetricsai. com/FileLibrary/Setup.zip for Auditmetrics.

When a business desires to obtain feedback from customers, it is usually disjointed from accounting information. The rdbms allows the linking of customer opinions with the accounting system. The Likert relationship links customer's opinions with their account activity. A table with a primary key is a unique identifier so that the table contains a list of unique entries that Access enforces transparent to the user. Relational databases use tables that are all connected to each other that are clearly visible and understandable.

The ingredients are also available to do a projection of sales by zip code. This will help in determining local market characteristics. If the customer table also has demographics data, such as age and gender, then there can also be an analysis of specific market demographic segments. These factors are essential ingredients in performing market research.

How to Use MS Access

MS Access is a powerful relational database that is easy to use. Below is a brief summary of how to get started in setting up a database. If you can put your relevant account data on an Excel file, Access can use it to load it on to a table. Below is a brief overview in setting up an Access database. It is beyond the scope of this book to go into great detail, but there are an ample number of web sites that can enhance knowledge in this very important topic.

MS Access is a powerful tool in creating and implementing a RDMS without the need to use code. It can be done by use of wizards or using the following steps.

- Determine the aim of your relational database.
- Define tables, fields, and primary key index.
- Determine the relationships between tables.

1. **Build a Microsoft Access database**

   - Start Microsoft Access, and then create a new blank database named, e.g., "MyAccess.mdb."
   - Create a new table named, e.g., "Sales Account" by following these steps:

     - Under Objects, click Tables.
     - Double-click Create table by using wizard.
     - Verify that a new table named "Sales Account" has been created.

2. **Create database relationships in Access**

   - Create a relationship—Create a relationship in an Access desktop database.
   - On the Database Tools tab, in the Relationships group, click Relationships.
   - If you haven't yet defined any relationships, the Show Table dialog box automatically appears.
   - Select one or more tables, and then click Add. After you have finished adding tables, click Close.

Once the tables for the rdbms are created, it is a simple matter to read in the appropriate data from external sources. This is where Access is very versatile. Data can be input using Excel and also .csv and tab-delimited text files which are the files of Auditmetrics.

Other possible data inputs:

- HTML Document
- XML files
- SQL Server
- Azure Database
- SharePoint List
- Outlook Folder

Some of Microsoft Access's most-used features:

- Importing and exporting data from Excel or other databases
- Creating forms for data entry or viewing
- Designing and running data retrieval queries
- Designing reports to be either printed or turned into a PDF
- Allowing users to interact with Access via SQL

Appendix 4: IRS Relative Precision Sliding Scale Confidence Interval

Today's companies are taking advantage of federal tax credits and deductions to generate much-needed cash. Claiming these credits and deductions often requires detailed review of large amounts of data. Statistical sampling is a useful tool in making a very large task manageable and is an even more cost-effective solution now under the statistical sampling guidance issued by the Internal Revenue Service.

Statistical sampling is used both by taxpayers and the IRS as a tool for the estimation and the examination of various numbers that appear on a tax return or claim for a refund. There has been a substantial rise in taxpayer use of statistical sampling since the IRS first issued its field directives.

Some of the areas in which sampling has been particularly helpful include:

- Producing estimates of qualified research expenditures for purposes of calculating the research credit.
- Estimating the amount of meal and entertainment expenses misclassified to accounts that are only 50% deductible.
- Generating figures for fixed-asset additions that can be depreciated over shorter useful lives.
- Reclassifying capitalized amounts as currently deductible repairs. Assisting in the determination of a taxpayer's domestic production deduction.
- The IRS has been using sampling for decades to examine the tax return supporting data that are too voluminous to approach with any other technique.

Over the years, as the IRS has become more involved with sampling concepts and the evaluation and use of statistical samples, it also has become more sophisticated in its approach. Since 2002, the IRS demonstrated its commitment to sampling by creating sampling coordinator positions. Sampling coordinators are responsible for communicating sampling's benefits, writing guidance and training materials, achieving nationwide consistency in application, and expanding the IRS's sampling knowledge and capabilities.

The IRS Shortcoming

It is the opinion of Auditmetrics early IRS statistical sampling original directive treated statistical relative precision ranges in a punitive manner. In the specific wording of the directive: "Taxpayer uses the least advantageous 95% one-sided confidence limit. The 'least advantageous' confidence limit is either the upper or lower limit that results in the least benefit to the taxpayer." Confidence limit is the range that an estimate from a sample will actually contain the true population value. Associated with confidence limit is the concept of precision. The IRS has introduced the concept of "relative precision" that is not the same as precision and margin of error as used by Auditmetrics and pollsters.

Table A4.1 IRS least advantageous estimate

2002 IRS Statistical Audit Directive		
Upper Bound	1,120,000	
Sample Estimate	1,000,000	
Lower Bound	**880,000**	**Taxpayer Haircut**

The least advantageous limit does have economic consequences when estimating a tax credit. If one is estimating a $1,000,000 research tax credit from a sample that has a relative precision of 12%, then the confidence interval would be between $1,120,000 and $880,000. Statistically the sample estimate and most probable value of the true book value for a tax credit is $1,000,000, but the IRS will allow only $880,000 which has a decided economic impact. Sometimes the use of the lower bound is also referred to as the taxpayer "taking a haircut."

The exhibit below outlines the impact using the least advantageous estimate to the taxpayer (Table A4.1).

Directive Updates

It should be pointed out that directives refer to "relative precision" which is calculated involving a one-sided 95% confidence interval. The description of relative precision from latest directive is:

Precision is a statistical measure describing how much an estimated value might vary. It is influenced by sample sizes, random selection, estimation method, and intrinsic variability within the population and sample data itself. It is that plus or minus amount that is reported with an estimate. For example, suppose total tax credits are estimated to be $1,000,000 plus or minus $80,000. Relative precision is a measure of how much an estimate may vary compared to its size. In this example, the relative precision is $80,000 ÷ $1,000,000 = 8%.

Relative precision is the measure of accuracy the IRS uses to assess taxpayer estimates. According to Rev. Proc. 2011-42, when the relative precision is below 10% (smaller is better), a sample estimate may be used without an adjustment due to estimate accuracy. However, when the relative precision exceeds 15% (larger is worse), the estimate is at the lower bound. Between 10% and 15%, there is a sliding scale adjustment.

There are other updated directives changes since the original directive concerning statistical sampling methods. The follow-up IRS publications on statistical sampling started the process of not placing limitations on the types of issues that may be addressed through sampling. Rather, the taxpayer only must demonstrate that the use of sampling is appropriate, i.e., by showing that the burden of evaluating the necessary data without sampling would be high and that other books and records do not independently exist that would better address the particular issue. In practice, the IRS typically does not challenge the use of statistical sampling per se. The directives now allow:

- Forgiveness of the sampling error "haircut" when the relative precision is 10% or less. For example, if the estimate produced by the sample is $1 million and the associated sampling error is $100,000 (10%) or less, the taxpayer may claim the full benefit of $1 million without any reduction or taxpayer haircut.
- Phase in of the sampling error haircut when the precision is greater than 10% and less than 15%, the taxpayer would be allowed to phase in sampling error haircut over the range of 10–15% on a sliding scale. Assume the sampling error is $120,000, or 12% of the $1 million estimate. Now IRS policy introduces a sliding scale, so the reduction to the tax benefit is not that large for sample estimates that just miss the 10% goal by a modest amount.
- **Inclusion of any certainty (detail) strata in the calculation of relative precision.**

The last bullet point is probably the most useful in estimating a tax credit with maximum benefit to the taxpayer.

These three scenarios indicate the IRS is liberalizing to use statistical sampling for a filing. In Table A4.2 there are three scenarios that rely heavily on adjustments of precision. A spreadsheet with an example is available on: https://auditmetricsai.com/FileLibrary/irsrelativeprecision.xlsx

As can be seen in scenario 2, the sliding scale allows for a tax credit of 952,000 just $48,000 below the maximum credit of $1,000,000.

Table A4.2 Three scenarios for estimating allowed credits based on relative precision

Scenario 1 if Rel. Precision <=10%	10%	
Upper Bound	1,100,000	
Sample Estimate (Allowed Credit)	**1,000,000**	**No Taxpayer Haircut**
Lower Bound	900,000	
Scenario 2 Rel. Precision >10% and <=15%	12%	
Upper Bound	1,120,000	
Sample Estimate	1,000,000	
Allowed Credit	**952,000**	**Sliding Scale Haircut**
Lower Bound	880,000	
Scenario 3 Rel. Precision >15%	15%	
Upper Bound	1,150,000	
Sample Estimate	1,000,000	
Lower Bound (Allowed Credit)	**850,000**	**Tax Payer Haircut**

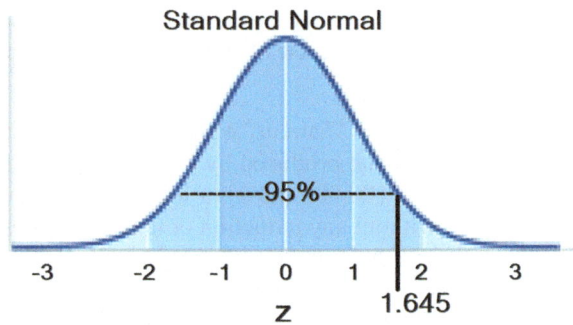

Fig. A4.1 Standard normal Z score for one-sided confidence interval

The IRS specification for the 95% confidence interval is "one sided" or a one tail confidence interval. The confidence interval formula is:

$$\tau\left(\text{tax credit}\right) \pm \mathbf{Z}_{.05} \times \sigma / \sqrt{\mathbf{n}}\left(\text{standard error}\right)$$

$$\text{Relative Precision} = \tau / \left(\mathbf{Z}_{.05} \times \sigma / \sqrt{\mathbf{n}}\right)$$

The standard normal distribution or Z score is 1.96 for a two-tailed test. The one-sided IRS sanctioned Z is 1.645 (Fig. A4.1).

Concluding Remarks

It is well-known that just keeping up with tax law and business day-to-day needs is time-consuming enough. But the trend in using statistical methods is on the rise and will only increase. For CPAs to ignore this trend as too lofty and not important will eventually result in a blind spot in maximizing benefits for their clients.

The IRS boosting the use of the statistical sample is part of a universal trend in using more and more sophisticated methods to enhance all types of analyses. Due to the explosion of information technology, what was once considered exotic is becoming commonplace. At one time our work in statistics revolved around complex statistical software, but now much of what was done with complex statistical software can be easily handled with Auditmetrics and Excel. That has caused a trend of converting a system of insider knowledge into a much more open-ended system with broad availability of knowledge.

The statistical sample in conjunction with market research can also expand business service line. But it is a tool where the IRS has set up very specific standards. For example, if a taxpayer wants to avoid a haircut, then make sure the relative precision of the sample is less than 10%.

Appendix 5: Tab-Delimited Auditmetrics Plug-In from MS Access

Microsoft Access is a versatile relational database system. A relational database (RDB) is a collective set of multiple datasets organized by tables. RDBs establish a well-defined relationship between database tables. Tables communicate and share information, which facilitates data search ability, organization, and reporting. The dataset for medical claims was derived from an Access-based claims processing system. The table selected is a subset of a much larger dataset with personal identifiers removed and non-standard medical procedure codes specifically set up for this exercise. Care was taken to assure compliance with the Health Insurance Portability and Accountability Act (HIPAA) privacy standards.

Paid Claim MS Access table design:

PaidClaims <--- Table Name	
Field Name	**Data Type**
Transaction_ID	Number
amount	Number
PlanName	Text
End_Coverage	Text
ServDate	Text
Error	Text
Service_Location	Text
Procedure Code	Text

This table was set up with Auditmetrics variables amount and Transaction_ID and Transaction_ID should be an "AutoNumber or a count of each transaction. Please note amount is also data type "Number," a more complete format listing for amount should be field size "double" and decimal places 2. If the data comes from a relational database, there may be a primary key that integrates several tables. This also should be included as a separate variable so that the sample can be integrated with other tables.

For the data needs of Auditmetrics what is also needed is absamt, absolute value for amount and sorted into ascending order. What also is required is the variable DataSet to identify this particular data for export. To do this you need to set up a Query.

If not comfortable in setting up an Access table and related query, an Internet search will reveal many sites that demonstrate how, at this juncture a table which contains the claims data. In order to add the two variables (absamt and Dataset), a query will be required. A query is a database object that creates a datasheet of specified records from one or more tables. A query allows an analysis of the data from different tables.

Claims Data Query

	Paid Claims Table		
Procedure_Code	absamt: Abs([amount])	DataSet: "HPlan1"	
PaidClaims			
	Ascending ←		

This exhibit is a segment of the query that contains all of the table variables plus two added variables required by Auditmetrics. Absamt is created with the script **absamt:Abs([amount]).**

In a query any text followed by a colon indicates that this is a new variable. After the colon the script indicates exactly the value of the new variable. In this instance it is the absolute value of the table variable amount. Also note that absamt is sorted in ascending order. <u>Be sure you use the query to sort absamt.</u> The second required variable is DataSet and designated by the script **DataSet:"HPlan1".** In this case the new variable has a single value HPlan1. What follows is a listing of the query.

Query1				
R	R	R	R	
Transaction_ID	amount	absamt	DataSet	Serv_Loc
96	9.95	9.95	HlthP1	HMO1
263	10	10	HlthP1	HMO1
260	10	10	HlthP1	HMO2
203	10	10	HlthP1	HMO1
284	10	10	HlthP1	HMO1
185	10	10	HlthP1	MD-Office
183	10	10	HlthP1	HMO2
235	10	10	HlthP1	MD-Office

The data is now in the form required by Auditmetrics is ready for download, so it can be loaded onto Auditmetrics. The "R" at the top indicates the specific variables required. Now the next step is to obtain a tab-delimited .txt file that can be read in by Auditmetrics.

I Select External Data & Text File

At this point go to the main Access menu bar at the top, and select "External Data" tab then "Export," and under file selections select text file. The first screen pop-up will ask some formatting questions you should ignore. Just move to the next screen by selecting OK. The next screen will ask what type file format you want to use. **Choose delimited**.

II Select Delimited then select Next.

O	Delimited - Comma or tab separate each field <--------				
O	Fixed Width—Fields aligned in Columns with Spaces Between				
Sample export format					
1	96	9.95	9.95	HlthP1	HMO1
2	98	10.95	10.95	HlthP1	HMO1

III Select the actual delimiter, in this example "Tab," and also include field names on first row and text qualifier as {none}, and then save the file with and extension of .txt for tab delimited or .csv for coma separated. **Be sure Text Qualifier is {none}.**

Choose The Tab delimiter that separates your fields

| Tab • | Semicolon 0 | Comma 0 | Space 0 | Other 0 |

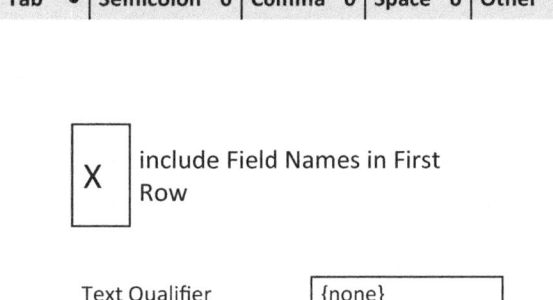

| X | include Field Names in First Row |

Text Qualifier | {none} |

Make sure you do these three selections, omitting any one can cause problems in the processing.

Appendix 6: Getting Started

1. Installing Auditmetrics

If you obtained from the Microsoft Store "Auditmetrics AI V6.5," it will be automatically loaded on the Windows operating system. If you downloaded from the Auditmetrics website, an icon will be placed on the home screen. If you are using an older legacy Windows computer and have problems, contact

Support@auditmetrics.com

If you are on a company server, it is best to contact IT to help with installation.

! For best performance of the AI Assistance software, it is best to first open Excel and leave it in the background. Then call up AuditmetricsAI.

2. Practice Data

Before downloading the data, create a folder such as Audit_Practice.

Practice Audit Data is available on the Internet Link: https://auditmetricsai.com/FileLibrary/Setup.zip

The link above will download a zip file. Click on the zip file, and extract all the files to your Aduit_Practice folder. The three files are:

Test_Population.xlsx—Original Excel file
Test_Population.txt—Converted from Excel to tab-delimited .txt
AudmetDB.accdb—Sample Access database

AuditResults.xlsx—Audit results file

Data Conversion to make it readable by Auditmetrics

- All Excel data can be easily converted Auditmetrics compatible by going to file, and select "Saving as" and then select "Text (Tab delimited) .txt".
- With a Microsoft Access dataset select at the top menu bar "External data," "text file," and "tab separator."

3. Auditmetrics Data Structure

! Auditmetrics requires four variables exactly spelled as below; if not present, an error message will be displayed. Letter case and order of the variables do not matter.

1. **Amount**—The transaction of interest in the analysis. Preferably, data collected should be total invoice value. For example, if the account to be sampled is for restaurant cash flow, it is best to record credit card total rather than each menu item.
2. **Absamt**—Absolute value of each transaction. This variable **must be sorted** in ascending order. This is to handle credits.
3. **Transaction_ID**—An identifier for each record; in this case, it is a record count.
4. **DataSet**—A name to identify this dataset.
5. **Primary Key- Optional** – If a dataset is from a relational database with a primary key that links the various data tables, it is prudent to include this variable in the audit population to be sampled.

Variables 1 to 4 must be in the data file and spelled exactly as above
! The variable Absamt must be sorted in ascending order.

Variable 5—If a dataset is from a relational database with a primary key that links various data tables, it is prudent to include it in the variables to be sampled.

Transaction_ID has value when the dataset is a merging of several data sources. For example, if you have two datasets of 1000 each and use MS Access to merge them into one file, then Transaction_ID 1 to 1000 is from the first dataset and 1001 to 2000 from the second dataset. Transaction_ID also has value when filing with the IRS. For more details, review *Appendix– Random Sampling and IRS Directives* in Book #1 in the "Small Business Power Series."

4. Let's Get Started
! At this Point it may be helpful to go to AuditmetricsAI.com and start the video to see an overview of the Auditmetrics system. That may all you need to get started.

1. The first step is to load tab-delimited Test_Population.txt and then select the command Button: "Potential Detail Cutoffs:

Σ Auditmetrics - AI Professional V6.5

For Help: info@auditmetrics.com

Detail _____

No. Strata   6 _____

Precision   .03 _____
(Margin of Error)

Efficiency _____

Total Sample _____

○ **Sample Size Excel File**

○ **Sample Validation Excel File**

| **Potential Detail Cutoffs** | ⇐

| **1. Sample Size Calculations** |

Detail Cutoff This is a dollar cutoff above which one reviews 100% of all transactions. A rule of thumb is that the detail stratum should represent approximately one-third of the total dollar volume, but Auditmetrics uses a more sophisticated statistical analytical approach.

! *Professional Version, Auditmetrics AI, has a button in the upper right-hand corner that allows the "Potential Detail Cutoff" button to also display Benford's Law first digit and second digit assessment. It is a useful forensic accounting tool to detect possible fraud.*

The Number of Strata The default value is 6 strata. A stratified random sample will yield more precise results than an unrestricted random sample of the same size.

Precision—The default precision or margin of error on the screen is 3%. If the audit is for a formal submission to the IRS or state revenue agency, 3% is the gold standard. If you are conducting an internal audit but just want to get a preliminary look at the data, then choose less precise values such as 5% or 7%. This will result in a smaller sample, thus reducing cost and quickly identifying potential data problems. Precision is the primary driver of sample size.

Efficiency Factor You should notice on the screen the measure "Efficiency Factor." The statistical meaning surrounding this measure is better left to the book, but the higher the efficiency, the better.

! Once the three inputs are decided, select the "1. Sample Size Calculations" tab.

5. **Step 1 Generating the Sample**

After deciding on the three inputs, it is time to generate the sample. We will use the inputs that were on the video: cutoff, 1600; precision, .03; and number strata, 5. You can not only generate the sample, but if you select the radio button "Sample Size Excel File," you will also generate an Excel spreadsheet that documents the sample design for future reference.

6. **Step 2 Validate the Sample**

If the sample specifications displayed on the screen are acceptable, the next step is to generate the sample. Select button **"2. Select Random Sample"** and **"3. Sample Validation."** These are all that are required, and two random sample files will be generated: "SampleData.txt" and "SampleData.csv." The .csv file is a comma-separated variable file, which is a text file that can be directly read into Excel and saved as an Excel workbook. The .txt file uses a tab as a variable separator. Both files will be saved in the same folder that had originally been set up for the audit population data.

Efficiency   0.72

Total Sample   1158

Sample Specs:

Freq.	Mean
10101	19.02
5776	91.67
2813	258.69
1614	548.66
888	1106.04
463	1880.24

● Sample Size Excel File

○ Sample Validation Excel File

Potential Detail Cutoffs

1. Sample Size Calculations

2. Select Random Sample

Acme Inc.

Strata:		Lower Range	Upper Range	Sample Size	Mean	Std. Dev.	Total Value	Pop. Total
0 -49.99	Stratum 1	0	50	67	$19.02	$14	$192,150	10,101
50-174.99	Stratum 2	50	175	124	$91.67	$43	$529,492	5,776
175-399.99	Stratum 3	175	400	138	$258.69	$95	$727,690	2,813
400-824.99	Stratum 4	400	825	161	$548.66	$185	$885,532	1,614
825-1600	Stratum 5	825	1,600	205	$1,106.04	$409	$982,161	888
> 1600 -Detail	Aud 100			463	$1,880.24	$0	$870,549	463
Totals Including Detail				1,158			$4,187,575	21,655
Sampling Efficiency	0.72							

This is the step to design the random sample. You can look in the folder you created for the sample file named SampleData.csv. This is a special text file; it is a comma-separated value (.CSV) file. The advantage of this special text file is that it can be read in by MS Excel spreadsheet. If the sample design is approved, then proceed with the audit. This text file can be "saved as" an actual spreadsheet file (.xlsx).

7. **Conduct the Audit and Document the Results**

The auditor can now use the file "SampleData.csv" to actual conduct the audit. Open the file "AuditResults.xlsx" that was downloaded to see how audit results are recorded. The SampleData.csv must pass the validation tests before it can be used.

Efficiency	0.72		N
			10101
			5776
Total Sample	1158		2813
			1614
			888
○ Sample Size Excel File			463
● Sample Validation Excel File	⇐		Sample S Validatio
Potential Detail Cutoffs			n
			67
			124
1. Sample Size Calculations			138
			161

Before proceeding with the audit, the generated sample must pass the validity test that are displayed on the screen:

Validation Tests Listed Below

n	Mean	SD.	Total $		
63	20.1	14.71	1266	ok	ok
61	72.49	43.7	4422	ok	ok
88	181.85	64.7	16002	ok	ok
90	369.14	96.52	33222	ok	ok
136	621.81	254.7	84566	ok	ok
152	1122.04	540.26	170550	ok	ok
463	1880.24	632.25	870549		

Validation #1– Observed precision under 0.03 no need to resample

Validation #2– Strata specific test passed.

This following Excel file section documents both the Audit Population (account being sampled) and that of the sample.

Acme Inc.

Strata	Population			Sample		
	Pop. Mean	Pop.Total Value	Pop. Freq.	Sample Mean	Sample Total Value	Sample Size
0 -49.99	$19.02	$192,150	10,101	$17.48	$1,171	67
50-174.99	$91.67	$529,492	5,776	$89.81	$11,136	124
175-399.99	$258.69	$727,690	2,813	$250.17	$34,523	138
400-824.99	$548.66	$885,532	1,614	$544.05	$87,592	161
825-1600	$1,106.04	$982,161	888	$1,108.75	$227,294	205
> 1600	$1,880.24	$870,549	463	$1,880.24	$870,549	463
Total (Excluding Detail)		**$3,317,026**			**$361,717**	**695**
Total (Including Detail)		$4,187,575			$1,232,266	1,158

In the next section is where the auditor records the results "Error Amount." Also displayed is the statistical validation test which indicates the 95% confidence interval test was passed for all sample strata.

Audit Results			Strata Validity Test		
Amount Error	Error Ratio	Pop. Est. Error Amt.	Lower 5% Alpha	Upper 5% Alpha	
$88.22	0.075	$14,471.99	13.85	$21.11	pass
$486.20	0.044	$23,117.15	83.06	$96.56	pass
$1,803.27	0.052	$38,010.00	233.48	$266.86	pass
$4,406.45	0.050	$44,548.18	515.40	$572.69	pass
$9,025.15	0.040	$38,998.57	1,065.84	$1,151.66	pass
$26,976.44	0.031	$26,976.44			
⇧	Auditor Enters Amt. Error				

Below is the audit result:

Error (Excluding Detail)	Std. Error ($)=	$26891			Sample Adj=	0.048
Error (Including Detail)					Overall Rate=	0.044
	-1 Std. Error	$132,255				
	Mid-Point	$159,146	Detail Error	$26,976		
	+1 Std. Error	$186,037				

It indicates that the statistical estimated dollar amount in error is $159,1460 with its band of standard error. The detail actual amount in error is $26,976.

Once a sample is drawn for an audit, it can be the basis to create datasets to forecast revenue and expenses. The key is to use the most powerful tools in building statistical projections, the principles of regression and correlation. Both of these statistical methods are also critical in linking the mathematical relationship between expense inputs and revenue outputs. Only when this relationship is statistically measured can a business consistently manage revenue and cost and forecast future economic performance.

Index

A
Accounts receivable, 3, 10, 17, 19, 34, 96, 111
AICPA Standards, 33
American Institute of Certified Public
    Accountants (AICPA), 4, 5, 33, 39, 86
Analysis of variance (ANOVA), 44, 93
Attribute, 10, 11, 33–35, 48, 64, 88, 96–98
Auditmetrics AI, 5, 8, 24, 36, 37, 42, 53, 54,
    84, 125, 127
Auditmetrics inputs, 37, 50, 114
Audit Model Online, 22
Audit population, 5, 6, 23–28, 35–37, 39–47,
    91, 101, 104, 114, 126, 128, 129
Audit report, 20

B
Benford formula, 53–55
Bias, 5, 8, 20, 53, 83, 96
Bimodal, 25, 45
Budget, 16, 69, 71, 73, 84

C
Cashflow, 3, 6, 10, 11, 13–20, 29, 49,
    72–74
Cashflow budget, 15–17
Central limit theorem (CLT), 39, 40, 42, 101–103
Certified Public Accountants (CPA), 4, 73
Confidence interval, 39–42, 44, 46–48, 61, 66,
    67, 96, 99, 102, 104–106, 119–122, 130
Correlation, 6, 7, 59, 61, 62, 69–71, 85, 107,
    108, 131
Curvilinear regression, 69

Customer profile, 81
Customer ratings, 9, 82, 84, 86, 115

D
Descriptive statistics, 4, 43, 87–93, 109
Detail strata, 23–26, 36, 37, 42, 43, 45, 47, 127
Direct labor, 70, 71
Dummy variables, 64, 65

E
Efficiency factor, 25, 37, 41–45, 127
Estimate, 5–8, 10, 14, 15, 24–28, 33–36,
    38–43, 46–49, 62, 66, 69, 94–98, 101,
    104, 119–121
Exponential growth, 74

F
Financial projections, 13, 59–66
Financing, 14, 17, 60, 71
Forecasting, 4, 7, 10, 14–16, 57, 59, 66, 69,
    71, 81, 84, 87
Forensic accounting, 53–56, 127

G
Goodness of fit, 53, 54, 73, 85

H
Hybrid approach, 34
Hypothesis testing, 102, 105

GPSR Compliance

The European Union's (EU) General Product Safety Regulation (GPSR) is a set of rules that requires consumer products to be safe and our obligations to ensure this.

If you have any concerns about our products, you can contact us on ProductSafety@springernature.com

In case Publisher is established outside the EU, the EU authorized representative is:

Springer Nature Customer Service Center GmbH
Europaplatz 3
69115 Heidelberg, Germany

The manufacturer's authorised representative in the EU is Springer
Nature Customer Service Centre GmbH, Europaplatz 3, 69115 Heidelberg,
Germany. If you have any concerns regarding our products, please
contact ProductSafety@springernature.com

Printed and bound by CPI Group (UK) Ltd, Croydon, CR0 4YY
23/04/2026
02095594-0015